T0302217

An Introduction to
Quantum Mechanics

An Introduction to Quantum Mechanics

From Facts to Formalism

Tilak Sinha

(Narasinha Dutt College, Howrah, India)

CRC Press
Taylor & Francis Group
Boca Raton   London   New York

CRC Press is an imprint of the
Taylor & Francis Group, an **informa** business

First edition published 2021
by CRC Press
6000 Broken Sound Parkway NW, Suite 300, Boca Raton, FL 33487-2742

and by CRC Press
2 Park Square, Milton Park, Abingdon, Oxon, OX14 4RN

CRC Press is an imprint of Taylor & Francis Group, LLC

Library of Congress Cataloging-in-Publication Data

ISBN: [978-0-367-54707-3] (hbk)
ISBN: [978-0-367-54729-5] (pbk)
ISBN: [978-1-003-09033-5] (ebk)

Latin Modern font
by KnowledgeWorks Global Ltd.

To my father,
for showing me how to value
only what is truly valuable

Contents

B Linear Algebra 221

Preface

Why another book on Quantum Mechanics? My excuse for indulging in this venture is not much different from my predecessors. I believe that there is something new and important that this book will offer and that it will be beneficial, at least, for some students. I must, however, admit that this book is not a substitute for a standard text book on Quantum Mechanics. It should rather be read as one long introductory chapter where the formal description of Quantum Mechanics is first introduced. It is a *constructive* approach to the inner product space formulation of Quantum Mechanics as opposed to *inductive* approaches based on intuitive associations or blatant *axiomatic* approaches aimed at the purely mathematically oriented.

The book is essentially aimed at undergraduate students taking their first course on Quantum Mechanics (although, I hope there will be something in it even for the experts). There is practically no prerequisite for this book. It is designed for anyone who is interested, and knows a little bit about matrix additions and multiplications. A philosophical taste and respect for logical purity is nevertheless assumed.

The main thesis of this book is the following:

*The entire mathematical fabric of quantum mechanics, built on inner product spaces and linear transformations, is essentially a consequence of the fact that **quantum probabilities interfere** - a fact that is encoded in the superposition principle.*

The formal description of any theory often obscures its empirical content. Quantum Mechanics is no exception. When a student first encounters the postulates of Quantum Mechanics (states are rays in a Hilbert space, observables are linear Hermitian operators, etc.), it is very hard for her to conceive how nature can suggest its nature in so unnatural a way. For example, the student has no clue what empirical fact is described by the statement that an observable is a linear Hermitian operator. Equipped with the postulates of Quantum Mechanics, a student can surely compute what Quantum Mechanics is designed to predict but the student usually has no idea how the postulates got cooked up. Despite all the skills that the student may have acquired in applying the postulates, they remain incomprehensible to her. A student needs to understand why a theory has been laid out in the way it has been. What other choices there were, and why the choices that were made, were actually made.

In this book I try to introduce quantum mechanics, first, in a simple, informal language - a language that would be used by an intelligent pedestrian who can perform experiments and draw inferences from them, but does not know about inner product spaces and linear transformations. Then, I show how one can recast this empirical content in the standard, formal language of quantum mechanics using inner product spaces and linear transformations. So, rather than starting by defining a quantum state as a normalized vector in a Hilbert space, and a quantum observable as a linear Hermitian operator acting on that Hilbert space, I start with more tangible definitions which conforms to the notions of state and observable that we have elsewhere in physics and in life. Then I show how the empirical laws of Quantum Mechanics refines these notions, eventually leading to the formal definitions.

I demonstrate, step by step the construction of the inner product space formulation from the empirical content of Quantum Mechanics to bring out the fact that *the inner product space formulation of Quantum Mechanics is a way of expressing the factual content of Quantum Mechanics and not the content itself.* It is important for a student to understand this difference between *content* and *language* so that she can appreciate the power and beauty of the *language* while being able to perceive the *content* independently of the language.

Of course, the most important point to see about a formulation is not *that it is possible* but to see *what it achieves.* The primary motivation for writing this book was to show what the inner product space formulation actually accomplishes. I have tried to demonstrate how this formulation leads to a pragmatic recipe for solving the typical problem that Quantum Mechanics was designed to solve by reducing the target questions to an eigenvalue problem.

The content and purpose of the book automatically addresses the issue of *measurement* in Quantum Mechanics. According to the standard interpretation of Quantum Mechanics, the so called Copenhagen Interpretation, *Quantum Mechanics is a theory of measurements.* Unfortunately, *measurement* is one of the most ill-understood and controversial terms in Quantum Mechanics. It is a source of immense worry for people who are concerned about the logical purity and foundations of Quantum Mechanics[1]. In the Copenhagen interpretation, measurement does not really *measure* a preexisting attribute of a quantum state of a system. It actually *prepares* a state which is *labeled* by what we *call* the outcome of the measurement. This view of measurement is indispensable if one wants to understand the Copenhagen interpretation - *even if it is only to find out what is wrong with it.* In this book I have made an aggressive effort to hammer this idea into the student's mind right from the outset. I will consider my effort amply rewarded if this one idea is well

---

[1]If John S. Bell had his way, he would have dispensed with the term *measurement* altogether from Quantum Mechanics and replaced it with something like *experiment.*

digested by the reader after reading this book. It is also one of the key themes of this work to demonstrate how the entire conceptual edifice of Quantum Mechanics is built from the single notion of measurement.

As much as I may have talked about the factual content of Quantum Mechanics, and despite the subtitle of the book, "From Facts to Formalism", I have purposely refrained from discussing actual physical experiments. Discussion of any real physical experiment is bound to raise subtle questions in the mind of the critical reader that eventually gets dangerously close to the delicate foundational issues of Quantum Mechanics (these are issues related to the *boundary of classical and quantum*). I wanted to spare the beginner from such worries. It is best that the student confronts such issues only after the basic structure of the subject is familiar. The strategy that I have followed is to keep the discussion of experiments sufficiently schematic so that the relevant features are pointed out and the disconcerting ones are never brought up.

Although, the main objective of this book has been to expose the facts that underpin the *formal description* of quantum mechanics and to make quantum mechanics seem *natural*, in going about this process, I have tried to provide the young student with a guideline on how to approach learning a new theory. Thus, somewhat unconventionally, I start the discourse with a discussion on the meaning of *theory* and its *formal description*. This allowed me to demonstrate, in concrete terms, how quantum mechanics is just another *theoretical framework* that conforms to the standard definition of a *theory* in the natural sciences.

The required linear algebra (which comprise the main mathematical background for quantum mechanics) is taught as we go along. In my experience of teaching Quantum Mechanics, I have found this strategy to be much more efficient compared to teaching linear algebra, all at one go, in a separate chapter as a mathematical prerequisite. However, to avoid lengthy digressions, most of the proofs of the theorems have been relegated to the appendix.

Since quantum mechanics is a probabilistic theory, the concept of probability is all over the place. In the appendix, we provide a very brief introduction to the bare essentials of probability. The last section talks about *probability measure*. Although this topic is more advanced than what we need in this book, we took the opportunity to discuss this to illustrate the idea of abstraction and generalization which is essential for the understanding of the content of the book.

The problems constitute an essential part of the book and the reader is urged to solve them meticulously. I have posted them at appropriate places inside the chapters and not necessarily at the end. Many of the important theorems have been included in the problems. Some standard content (e.g., parity operator, Heisenberg picture representation), customarily included within the text, have also been relegated to the problems because they did not belong to the main line of development of this book. However, such problems have often been broken up into parts and provided with detailed hints so that by working sequentially through the parts the student will be able to acquire a basic understanding of the topic. The level of the problems is, at times, slightly higher than that of the book. While the main text requires practically no mathematical background (certainly not beyond elementary high-school mathematics), so that the main content is accessible to any (motivated) pedestrian, this is not true for the problems. The problems will often require some hard work. The book has been designed such that, together with the problems, it can serve as a standard text for a one semester undergraduate course in Quantum Mechanics.

The sections and even the equations in this book have not been numbered. This has been done deliberately so that the experience of reading this book is like listening to a lecture. Since the book is predominantly about conveying the motivation and spirit behind a formulation, it was important to adhere to this style to spare the reader from keeping a mental tab open in various pages which would compromise focusing on the main issues.

I am deeply indebted to all my students for their enthusiasm, questions and comments but above all for being the reason for this book.

I would like to express my gratitude to my teachers, and the authors whose books shaped my own understanding of Quantum Mechanics.

It is impossible to thank my wife Tanaya enough for putting up with all the hurdles that this project has been through. However, I would like to thank her, in particular, for carefully reading the manuscript, and for making critical suggestions. I would also like to thank her for choosing to use the content of this book for teaching Quantum Mechanics to her undergraduate students at St. Xavier's College, Kolkata, for the last several years, and for assimilating invaluable feedback from them. Indeed, it has been the most worthwhile collaboration of our life as physics teachers.

I would also like to thank my former colleague and friend Dr. Saurav Samanta for his valuable comments and encouragement.

Finally, I would like to thank the people at the CRC Press for their continuous help and support in making this project possible.

Tilak Sinha

Prologue

Quantum Mechanics (QM) is a theoretical framework that describes physics of the microscopic world (loosely, systems of atomic dimensions and smaller). Some people, probably most, believe that QM is a fundamental theory, and it should apply to everything - across all dimensions and not just to the microscopic. They say that, after all, large scale macroscopic objects are in the end made up of microscopic constituents; it is due to the complexity of large systems that the quantum signatures are somehow lost in macroscopic phenomena. What we know as classical physics, that applies to large scale phenomena, must emerge from an underlying quantum machinery. Be that as it may, whether or not QM is applicable to large systems, this much is certainly true that classical physics is not applicable to microscopic phenomena, and QM becomes *indispensable* in that regime. Even within this restricted domain of applicability, QM embodies an enormous volume of our experience. For example, it explains why water is the way it is, how long the sun will shine or why you are not falling through the floor as you are reading this. The importance of QM can therefore hardly be exaggerated. It is indeed, a corner stone of modern day physics.

QM can initially be a difficult subject to learn. Part of the difficulty is attributable to the fact that its applications relate to small-scale phenomena that are not directly perceptible to our biological senses. For example, water is the way it is (it has a certain density, specific heat, boiling point, chemical properties, etc.) because QM requires it to be that way, but we do not actually see the laws of QM at work that makes water behave the way it does. For large scale phenomena this is not the case. For instance, we explicitly see Newton's laws at work (at least, qualitatively) when we watch a tennis ball being hit.

The other, more serious reason that makes learning QM difficult is the fact that, as a theory, QM adopts a philosophy that is completely different from its classical counterparts. This requires us to dismiss our usual pattern of thinking and embrace a new one. For example in Newtonian mechanics (more particularly according to the, so called, *mechanistic view of the world*), one believes that all matter is essentially an assembly of particles, and once we know the positions of the particles as a function of time we know everything that is there to know. Every question that we ask about a system can be translated into a question involving the positions of the constituent particles of the system as a function of time. Thus, Newtonian mechanics is essentially aimed at solving the question: *"how do positions of particles change with time?"* QM, on the contrary is not obsessed with systems of particles. In fact, QM *per se* does not even talk about the constitution of systems that it tries to describe[2]. It is rather a general theoretical framework for describing measurements made on an arbitrary system. The question that QM hopes to solve is "at any given time, *what are the possibilities in the outcome of an arbitrary measurement made on a system, and what are the probabilities of the different possibilities?*". So, we see that the philosophy of QM departs from the philosophy of classical mechanics in at least two main points:

1. QM is a probabilistic theory as opposed to its deterministic, classical counterpart.

2. In QM, *position* loses its position of fundamental importance.

Owing to the difficulties discussed above, QM has earned a reputation of being a strange and mysterious theory. While it is alright to be shocked by it, or be excited about it, we must remember that *to understand something is to be able to see it as natural, and not as mysterious or magical.* Our emphasis will therefore be to try to appreciate how *natural* QM is as a natural science, and not on how *peculiar* it is.

---

[2]This does not mean that QM *cannot* describe composite systems *constituted* of smaller subsystems. It only means that QM does not restrict the nature of such subsystems. In particular, it does not force us to assume that all systems are constituted of some elementary system of a particular kind (which may be called a particle) that must have an attribute called position.

Chapter 1

Theoretical Framework - A Working Definition

The principal objective of this book is to make quantum mechanics seem *natural*. We wish to demonstrate how quantum mechanics is just another *theoretical framework* that conforms to the standard definition of a *theory* in the natural sciences, and it is our purpose to expose the facts that underpin the essential fabric of the *formal description* of quantum mechanics. Obviously, before we embark on this task, we must be clear in our minds what we mean by a *theoretical framework* and its *formal description*. So, in this opening chapter, we begin our journey by looking into the meaning of these terms.

What Is a Theory?

In the natural sciences, a theory is defined as follows:

A theory is a logical framework comprising some assumption(s) from which one can make some logical prediction(s) that, in principle, should be falsifiable by experiment[1].

The assumptions of the theory are usually motivated by factual observations[2]. They go by various names such as laws, principles, axioms, and postulates. The various names are used in slightly different contexts, but they essentially play the same role in the logical system. Here is a simple example of a theory. It makes the following two assumptions:

A1 Red apples are sweet.
A2 Sweet apples are expensive.

Since we want to discuss this *toy* theory for a while, we will give it a name.

---

[1]The *experiment* need not necessarily be *performed* by humans. It could well be some natural phenomenon that humans can *observe*.

[2]Quite often, however, the assumptions are educated guesses which do not lend themselves to a direct observation. It is only their implications that can be subjected to observation in such cases.

We will call it *"a theory on the price of apples"* or ATOPA in short. From the two assumptions, A1 and A2, we can immediately deduce an implication:

T1   Red apples are expensive.

This, of course, is certainly falsifiable by experiment. All we need to do is go about asking the price of red apples in every market that we come across.

It is important to note that the statement "If A1 and A2 is true, then T1 is true", is always correct (true) irrespective of the truth of A1, A2 or T1. Whether or not A1, A2 or T1 is actually true in real life is a separate issue. If T1 turns out to be false, then we say that ATOPA has been disproved or falsified by observation. In this case at least one of the assumptions of ATOPA (i.e., either A1 or A2) must be false.

It should be emphasized that *a theory can never be proved; it can only be disproved*. Every time a theoretical prediction fits observed data, it merely increases our confidence in the theory. No matter how many times this agreement (between observed data and theoretical prediction) occurs, the theory is not proved. The possibility of a hitherto unperformed experiment which does not agree with the theoretical prediction can never be ruled out. Moreover, even if there are no disagreements with observed data, it is always possible to imagine the existence of a yet undiscovered theory which makes the same predictions so that it is impossible to claim that a given theory is the *correct* one. On the contrary, a single contradiction of theoretical prediction with observation immediately *falsifies* the theory in question.

Formal Description of a Theory

To put the discussion into context, let us start with an anecdote.

The deaf composer

In a town there lived a music composer who had conceived a theory on the *musicality* of music. He was extraordinarily gifted and could actually *recognize* the patterns that made some sequence of notes *musical*, some *discordant*, and some *in between*. The composer wanted to develop and propagate his theory. He was so ambitious in his objective that not only did he want to specify a rule that would tell one which pieces will be musical and which not, he also wanted to invent a formula that would *quantify* the musical quality of a piece. In fact, he even wanted to develop a recipe that would *produce* good music!

Unfortunately, the composer lived in an era when musical notation was not yet invented. So the only way he could communicate his ideas was through actual demonstration. His ideas, naturally, did not propagate very far. Firstly, they were limited to his students and friends. Moreover, how much of his ideas would be grasped in a demonstration was crucially dependent on the perceptibility and sensitivity of the audience. The communication of the ideas through his followers was even more vulnerable to distortion because it was also heavily dependent on the quality of demonstration, in addition to the sensitivity and perceptibility of the recipients. So each layer of prospective followers in the chain was receiving an increasingly deviated version of the original theory. As was to be expected, the distorted versions of the theory were more and more, less and less impressive. The future of the theory looked grim! Our composer became very depressed at the inevitable fate of the science that he had given birth to.

At this point a mathematician in the town, who had only a modest appreciation of music, developed a scheme by which music could be represented in a visual form as a script, much the same way that spoken words are scripted using alphabets and punctuation. This was the birth of the musical notation!

Our composer immediately realized that this could save his theory. So he took the plunge and invested all his time and energy into developing and writing up his ideas in the newly invented language. He wrote down all the facets and features of his musical theory, starting from the recipe of *identifying* a truly musical piece to the formula for *creating* one. The results were exactly as he had expected. In a few years, there was an immense following of his theory. With the distortions hugely minimized, people really began to appreciate the true content and merit of his work. *It was all because the visual, scripted representation provided a much more objective character to his theory*; it was now much less open to subjective, personal interpretations.

Our composer quickly became a famous man. He came to be recognized as a father figure of the science of music. However, it was only after many years that he received his highest accolade. It was at a concert where he was invited as the chief guest. The showstopper of the evening was a piece composed by a young man about whom our composer had not even heard of before. The performance was so overwhelming that it left the audience speechless for some time. Then the entire audience rose to give the young composer a standing ovation. As he was being applauded by everyone, the anchor walked up to the middle of the stage and announced: "If you are all astounded by what you heard just now, I have more astonishment in stock for you. Our composer can't hear a single clap that you are showering on him. He was born deaf. He cultivated the art and skill of making music based on the theory and method that was developed by our chief guest tonight!".

This was a fictional anecdote designed to demonstrate the purpose of what is known as a *formal description*. A formal description is to its informal counterpart what the scripted version of the musical theory was to its original version (that was almost going to get obliterated). The whole purpose of a formal description is to make it so completely free of subjective human interpretations that even a *machine*, which is programmed to read the language, will be equipped to use it in spite of the fact that it has no *understanding* of the underlying *reality*. Let us then try to illustrate, more concretely, what we mean by a formal description of a theory in the natural sciences.

Formal description in the natural sciences

In the natural sciences, a theory is essentially a logical scheme to describe our experience. It establishes a connection between some set of our experience (usually called phenomena) and some logical system. The more mature the state of the science, the clearer the connection. In the initial stages of the formulation of a theory, the description (connection of the phenomena to the logical system) usually suffers from having ambiguities and, consequently, interpretations can be subjective. The ambiguities may be so subtle that they may not cause practical problems for a long time until some day one finds oneself confronting a situation that demands a resolution of the ambiguity. For instance, in the context of our theory ATOPA, it could so happen that one day someone comes across a breed of apples, which when ripe, become *red only on one side* (may be, around the stem of the apple). This would naturally lead to an ambiguous interpretation of *red apples*. To resolve this, one has to decide, precisely, to what extent the apple must be red (*fraction* of the surface that is red, and also perhaps the wavelength specifying the *redness*) in order that it may be called a *red apple*. In the absence of such a precise definition, ATOPA may be a correct theory in some interpretation (of a red apple) and false in another.

As our understanding of the underlying phenomena improves, ambiguities are resolved and the true shape of things emerge. When the concepts and assumptions necessary for a theory are precisely understood, one may try to formulate the theory completely in terms of well-defined mathematical entities and relationships. We call this a *formal* description of the theory.

Let us try to see what a formal description of a theory may look like. We shall, once again, use as an example the theory ATOPA, which we had described *informally* in the last section.

ATOPA (Formal Description):

- There exist three sets A, C and T whose members are respectively called "apples", "colours" and "tastes".

- There exist three function c, t and p on A, having C, T and $\mathbb{R}$ as their respective codomains:

$$c : A \longrightarrow C$$
$$t : A \longrightarrow T$$
$$p : A \longrightarrow \mathbb{R}$$

Here $\mathbb{R}$ is the set of real numbers.

The functions c, t and p obey the following conditions[3]:

1. There exists a colour "red" $\in C$ and a taste "sweet" $\in T$ such that for all apples $a \in A$

$$(c(a) = \text{red}) \quad \Longrightarrow \quad (t(a) = \text{sweet})$$

2. There exists a real number $e \in \mathbb{R}$ such that for all apples $a \in A$

$$(t(a) = \text{sweet}) \quad \Longrightarrow \quad (p(a) \geq e)$$

A formal description of a theory is sometimes called an *axiomatic description*. The assumptions of the formally described theory are generally referred to as *axioms* or *postulates*. Thus, in the above example, the assertions 1 and 2 may be called the axioms (or postulates) of ATOPA. It is important to understand that once an axiomatic formulation has been laid down, it becomes a piece of mathematics and can exist in its own right, independently of the phenomena it was designed to describe. Thus,

consequences of the theory ATOPA can be derived as theorems following from the axioms by a colour blind, taste blind person who has never heard of apples.

Since a formal description is independent of individual perceptions, it is free of ambiguities arising out of subjective interpretations[4]. Having said that, one must understand that for an axiomatic description to be of any practical value, it must carry with it a set of interpretative rules that establishes its connection to reality. For our theory ATOPA, the rules, of course, are obvious. Real apples are identified with elements of the set A, real colours with the elements of C and real tastes with the elements of T. If for an apple a we have $c(a) = \text{red}$, it would mean the apple *is* red; if $t(a) = \text{sweet}$, the apple *is* sweet; if $p(a) \geq e$, the apple *is* expensive. A well-formulated theory in the natural sciences should be considered as the union of some set of axioms and a suitable body of interpretative rules.

Abstraction and Generalization

An axiomatic formulation has to start by identifying appropriate mathematical entities that can be used to describe the different physical notions

---

[3] For those unfamiliar the symbols ' $\Longrightarrow$ ' and '$\in$' stand for *"implies"* and *"belongs to"*, respectively.

[4] At a deeper level, it can be argued that even formal descriptions are subjective in the last analysis, but we do not intend to delve deeper into this philosophical intricacy at this level.

that the theory intends to talk about. It is done through a process called *abstraction and generalization* that we shall discuss in this section.

Imagine a group of scientists, studying the mechanism of movements of animals with a purpose of implementing them in robots. After studying quite a handful of them (say, dogs, horses, tigers, kangaroos, humans and spiders), they observe a pattern. They see that the movement of animals that have four legs have a lot in common which can be entirely attributed to the fact that they have four legs. The scientists then decide to *define* an *object* that has four legs and *call* it a *four-legged animal*. This four-legged animal is, of course, not any real animal. It is a *hypothetical, faceless creature* defined only by the property that it has four legs. The scientists then get down to study all the logical consequences of having four legs in the movement patterns of the four-legged animal. This enables them to study many of the relevant properties of the movements of a large set of animals, *collectively*. The set of properties need not necessarily constitute the full characterization of the movement of some particular four-legged animal (which could be attributed to something else, for example, having a particular kind of tail). Nevertheless, everyone would agree that defining the four-legged animal provides for an efficient and economical method for the study. Apart from economy, there is another advantage of studying the four-legged animal: a great deal would be already known about all animals with four legs that would be encountered in future.

This method that we described using an analogy is known as *abstraction and generalization*. The identification and extraction of the relevant properties (e.g., "animal has four legs") is called *abstraction*. Defining a *general* class that is assumed to have the abstracted properties is called *generalization*. It is to be noted that this general class is not restricted to the union of the sets from which the properties are originally extracted.

This ubiquitous mechanism permeates all our understanding. It is something that we consciously or unconsciously do all the time. If we did not, then we would not have been able to walk on a surface on which we had not walked before. We can do it because our brain abstracts the relevant properties of surfaces (with regard to balancing, etc.), from a finite sample of them that we are exposed to in our childhood. It then sets up a set of rules of muscle coordination that would keep us from falling when we walk on some new surface, which the brain assumes to have the same relevant properties. One abstracts those properties from experience which are thought to be relevant for the question(s) of interest and internally forms a general class whose members are assumed to obey these properties. At a higher cognitive level, it is also the underlying mechanism by which concepts are formed and theories are made through *inductive reasoning*. Indeed, according to some schools of thought, *intelligence* is viewed and measured as the ability to see abstractions and generalizations[5].

---

[5]We shall make explicit use of this process in the third chapter when we introduce *abstract vectors*. Another illustration can be found in the last section of appendix 'A' that discusses the axiomatic description of probability.

A word of caution! In constructing a formal description of a theory, the process of abstraction and generalization can often be a rather slippery business. It is because, from a logical point of view, the mathematical entities become the definitions of the physical concepts they describe.

If the relevant physical idea is not captured in its entirety by the definition, it could lead to mathematically correct solutions which are completely useless from a physical point of view.

The following popular anecdote will illustrate the point[6]. A mathematician was once asked, "how do you catch a tiger?" To this, the mathematician replied, "before I start looking for a solution, I would need a precise definition for catching a tiger". This was, of course, not easy to provide. After a lot of thought, someone came up with the following definition: "by catching a tiger one means, isolation of the tiger from the person who is catching the tiger by means of a cage". When this was communicated to the mathematician, he instantly responded "that's trivial: you get inside the cage, then the tiger is caught - by definition".

Philosophy and Fundamental Theorem

Construction of a theory is guided by the following objectives:

1. Formulation of the *target question* that the theory aims to solve[7].

2. Construction of an *algorithm* that determines the answer to the target question.

The target question embodies the *philosophy* of the theory. For example, in classical mechanics, the target question is "how do the positions of particles change with time?" The reason why this is considered to be the target question is encapsulated in the mechanistic view of the world: *The universe is largely empty space inhabited by a (sparse) distribution of matter. All matter is essentially an assembly of elementary building blocks called particles. The particles have an attribute called position which is a function of another entity called time. Once we know how positions of particles change with time, we know everything that we could possibly want to know within the scope of*

---

[6]This well-known *joke* circulates in academic circles in many variants in various contexts. The earliest reference of the idea known to the author is: Ya Khurgin. *Did you say mathematics?* Mir Publishers. Moscow, 1974.

[7]It is quite possible to have more than one target question in a theory. However, having many target questions makes a theory practically and aesthetically less appealing.

classical mechanics. It is important to understand that the construction of a theory does not start with the formulation of the target question. In order to come to the point where the target question can be framed, one has to already have some assumptions in place that paints the underlying picture. In classical mechanics, for instance, the target question makes sense because we assume that there exists an entity called particle which has a property called position, which is a function of time. How we choose to *see* the things that the theory talks about is what we mean by the philosophy of the theory[8].

Once the target question has been framed, the next step is to look for some clues that would enable one to make some more (usually, not so obvious) assumptions, which together with the other assumptions already laid out, lead to an *algorithm* for obtaining the answer to the target question[9]. In Newtonian mechanics, these assumptions are the famous Newton's laws of motion. The algorithm is a theorem that is derived from the assumptions of the theory. In a real-life theory, this will typically be based on several theorems that orchestrate to yield one grand theorem, which is the statement of the algorithm. We can refer to this theorem as the *fundamental theorem* of the theory.

Thus, to summarize, *there are* two *main objectives that drive the construction of a theory*: *adopting a* **philosophy** *(embodied by the target question) and deducing the* **fundamental theorem** *(which provides the algorithm for solving the target question)*.

Theory vs Theoretical Framework

In our discourse, it will be important to understand the distinction between the terms *"theory"* and *"theoretical framework"*. The term *theory* is used when we talk about the description of a particular phenomenon while *theoretical framework* refers to a general logical structure shared by a class of theories. *The theories under a theoretical framework share the same philosophy, and the same general algorithm (validated by the fundamental theorem) that applies to all of them.* For example, Newtonian mechanics is a theoretical framework that can be used to write down a theory that describes the mo-

---

[8]The etymology of the word *"philosophy"* is "love of wisdom". It comes from the Greek words *"philos"* meaning "to love", and *"sophie"* meaning "wisdom". In Sanskrit, philosophy is called *"darshan."* Translated literally, it means "to see". In our context, this *sense* of the word is more appropriate. Philosophy of a theory is how it chooses to *see* things.

[9]An *algorithm* is an instruction comprising a sequence of well-defined steps to perform some specific task. Every step in an algorithm, by definition, must be *executable*. Here, we use the term in a slightly *loose* sense. The *"algorithm"* that we refer to can involve steps (such as solving a differential equation) which may not be *exactly* solvable in all possible cases. What we mean is: *an algorithm is a sequence of steps such that when all steps are executable, it leads to a solution of the target question.*

tion of a planet moving around a star, or a theory that describes the motion of a charged particle moving in a magnetic field. For both of these *theories*, Newtonian mechanics is the underlying *theoretical framework*. In the framework of Newtonian mechanics, one has to specify (i) the intrinsic properties of a system (e.g., mass of particles), and (ii) the *force law(s)*, as theoretical input[10]. Once these inputs have been specified, we have *one* theory belonging to the family of theories in the Newtonian mechanics framework.

Before moving on, we would like to mention, however, that often for the sake of brevity, if there is no chance of confusion, a "theoretical framework" is referred to, simply, as a "theory".

---

[10]In the example theories cited, the law of gravitation and the laws of electromagnetism serve to provide the force laws.

Chapter 2

The Empirical Basis of Quantum Mechanics

In this chapter we will try to introduce the basic empirical content of QM. However, rather outrageously, we shall do it without discussing any actual physical experiment. Instead, we will use cartoons to describe the general results that have been extracted from such experiments. Even these results shall not be presented in their full glory. We will set up a simplified scenario which will allow us to quickly weave the basic mathematical fabric of QM.

We will start with a metaphorical description designed to illustrate the typical setting of a quantum mechanical problem. This will be followed up with a more physical illustration of the core aspect of quantum measurement before we get to the actual business of describing QM.

Professor Funnyman's Ghosts[1]

Professor Funnyman was a staunch believer in ghosts. So much so that he took it upon himself to study them in his laboratory. The ghosts, of course, were not perceptible to him any more than they were to others. So, to his critics, professor Funnyman had a real hard time *defining* the systems he claimed he was working on. Nevertheless, he devised experiments that he *believed* measured various properties of the ghosts, like their colour, odour etc. Strangely, that led to a solution to his problem. Here is how.

The results of professor Funnyman's experiments (see Figures 2.1–2.4) could roughly be summarized as follows.

1. *A specific measurement on a ghost, designed to measure a specific ghost property, always produced one or the other of a specific set of values.* For example, upon measurement of colour (which means execution of some well defined actions involving a particular apparatus that professor Funnyman believed will determine the colour of the ghost) professor Funnyman always found red or green (which means there was a window in his colour apparatus which glowed red or green) and never anything

---

[1]For those who are already familiar with QM, this metaphor will apply to QM without degeneracy.

else. Similarly, measurement of odour always yielded coffee or lemon (which means there was a hole in the odour apparatus that one could sniff at, and it smelt either of coffee or of lemon) and never anything else.

2. *When two successive measurements were performed, there was always a definite probability for every possible sequence of outcomes and a given delay between the measurements*[2]. For example, if successive measurements of colour and odour were made, separated by a fixed interval (say five minutes), there was always a definite probability for getting coffee after red or lemon after green. Thus, it was always the *last* measurement result that determined the probability of the outcomes of *all* the future measurements. Measurements made prior to the last measurement had no bearing on outcomes of future measurements because definite probabilities could be assigned *only* to *pairs* of outcomes resulting from a sequence of *two* successive measurements. So, according to professor Funnyman, *ghosts have a very short memory - they only remember the last measurement made on them.*

3. *If the same measurement was repeated instantaneously, the outcome of the first measurement was always reproduced in the second.* For example, when professor Funnyman measured colour twice in quick succession (i.e., with a vanishingly small delay), then either both outcomes were red or both outcomes were green (i.e., probability of getting red after red or green after green was unity, while red after green or green after red was zero). However, if a measurement of colour produced red and an immediate measurement of odour produced lemon, then another instantaneous measurement of colour was no longer guaranteed to yield red. There was a definite probability for getting red, lemon and then red again (which is the product of the probabilities of getting lemon after red and red after lemon) but that probability was not unity anymore. This was a curiously strange feature of ghosts. But then, they were ghosts after all. Also, if there was a delay between measurements, it was in general no longer certain that the outcomes of identical measurements would be identical. If one made a colour measurement that yielded red and then repeated the measurement after waiting for five minutes, the outcome would not be red with certainty.

The results of his experiments led professor Funnyman to the most important realization of his life. He understood which questions were meaningful and which were not. He understood that he was only *obliged* to answer questions about what can be *observed.* If the possible answers to some question was not experimentally verifiable then he did not need to answer such a question[3].

---

[2]Since quantum mechanics is fundamentally a probabilistic theory, it is important to have a reasonably well-formed idea about probability before proceeding with the content of this book. We have provided a brief description of probability in the appendix for those who are not familiar.

[3]This, of course, does not mean that it is forbidden to have in a theory concepts that are

FIGURE 2.1: Three consecutive, instantaneous measurements: colour-odour-taste. With fixed delay between measurements, the probability P is associated with a *pair* of *consecutive* outcomes. Thus, P [red → coffee → bitter] = P [green → coffee → bitter] = P [coffee → bitter].

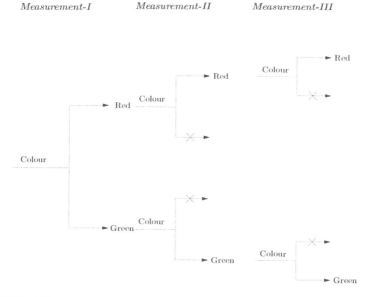

FIGURE 2.2: Repeating the colour measurement, instantaneously, reproduces the same colour. Thus, P [red → red] = 1 = P [green → green].

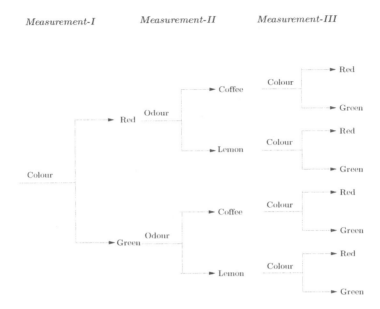

FIGURE 2.3: Previous colour is not reproduced because of intermediate odour measurement. Thus, P [red → lemon → red] ≠ 1 ≠ P [green → lemon → green].

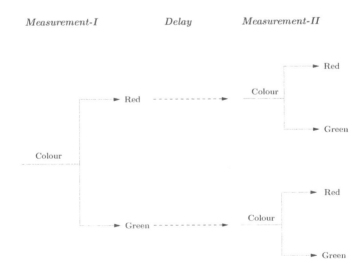

FIGURE 2.4: Consecutive colour measurements may not produce the same colour if there is a delay between measurments. Thus, P [red --→ red] ≠ 1 ≠ P [green --→ green].

The important thing was that, for the system that he was studying, there were some questions that could be experimentally answered. To wit *"if a measurement of some observable was performed, what were the possible outcomes and what were the probabilities of the different possibilities?"*.

Now for the definition of ghosts, professor Funnyman decided that instead of defining ghosts first, he would start by defining *"ghost-states"*. He emphasized that at this point a *ghost-state* was not to be understood as the state of the ghost but simply as a whole entity that he chose to call a ghost-state. According to him a *ghost-state was to be characterized by the outcome of the last measurement*. Usually by *"state"* of a system we mean some minimal information pertaining to the system from which one can make all possible predictions within the scope of the theory. For the theory that professor Funnyman had in mind, the predictions could only concern the probabilities of outcomes of future measurements. Since professor Funnyman had found that once a ghost-measurement was performed and an outcome recorded, it determines the probability for every outcome of every possible subsequent measurement, it was natural to use the last measurement result to characterize *state*. Thus, *every measurement outcome of every observable would characterize a possible ghost-state*. Professor Funnyman's definition of (ghost) states implies that *for ghosts, a measurement ought to be considered as a state preparation process as opposed to a process that measures some preexisting property of a system*. Thus, "red" would label a ghost-state that is *prepared* by the colour measuring process when the outcome is red, and not as the colour of the ghost *measured* by a colour measurement. This, of course, was also justified since professor Funnyman did not have a prior definition of a *ghost* so it would make no sense to ascribe a property to it. Moreover, according to professor Funnyman, one could not be sure, for example, whether the glow that one sees through the window of the colour measuring instrument is really the colour of the *ghost* (if one insists on its existence) or a result of the interaction of the ghost with the measurement apparatus. All that one could tell, for sure, was that a colour, say red, was an outcome of colour measurement that could be reproduced if the measurement was repeated instantly, and which determines the probabilities of outcomes of all future measurements[4].

Having defined the ghost-states it was now easy for professor Funnyman to define his system - the ghost: *A ghost was defined as something that could exist in one of the ghost-states*. This was based on the simple idea that if a ghost exists, it must exist in one of its states. So, *the ghost was characterized by the universal set of ghost-states, which was essentially the set of outcomes of all possible ghost-measurements*. Thus, to summarize: **ghosts were characterized by the totality of ghost-states, the ghost-states were specified by**

---

not accessible to experiments. A real theory can and usually does contain such concepts. The point is that it is not *compelling* on a theory to contain a concept, or answer a question that is not accessible to observation.

[4]However, there is no harm in using a vocabulary inspired by *imagining* that red *was* the colour of the ghost as long as one is aware of what it really means.

the way they were prepared, and the state preparation process was identified with measurement process.

From here on, professor Funnyman did not have to look back. He happily continued with his research on ghosts with full commitment and zeal. Along the way professor Funnyman attracted many followers and they all joined hands in an effort to construct a theoretical framework that would answer the basic questions about ghosts.

If we come to think of it, the microscopic systems (that QM was designed to describe) like atoms or nuclear particles are very similar to professor Funnyman's ghosts. They are also not directly perceptible by our biological senses. We *see* them only indirectly with the help of some measurement apparatus. But all measurements on such systems come with associated interpretations based on our psychological notion of the corresponding observable which is formed entirely from macroscopic experience. It is then more appropriate to *define* the properties of such systems by the way they are measured rather than regarding measurement as a means of determining the properties of such systems.

An Experiment with Bullets

Here we try to give a slightly more realistic (physical!) example to illustrate the main point that we tried to make in the preceding paragraph.

Imagine an experiment with *invisible* bullets (perhaps because they are very small) that is designed to measure the velocity of the bullets. We *assume* that the bullets are generated and ejected from a gun by some mechanism (just as electrons can be liberated from some metal surfaces by heating them). We also assume that there is a provision in the gun which allows the velocity of the bullets to be continuously varied (just as electrons can be collimated and accelerated by a continuously varying potential). For example, we can imagine a trigger-spring (obeying Hooke's law) which can be stretched continuously. All these assumptions are based on expectations which has its roots in classical physics. The bullets finally impinge on the surface of a specially designed material and create *dents*. The velocity of a bullet is then inferred from the depth of the dent based on some classical theory once again.

The exact size of the dent will depend on the property of the material of the surface which receives the bullet and the gun that determines its speed. These parameters may not be completely controllable or even precisely known[5]. Nevertheless, this much is certainly expected that the inferred velocities should display properties consistent with the physics that has gone into the design of

---

[5]When this is the case, one might need to suitably calibrate the apparatus with known velocities first.

the apparatus. For instance, it is certainly expected that the observed velocities will display a continuous spectrum.

Now let us imagine that, for some reason, the outcome in the measurement exhibits some strange feature. For example, let us suppose that the observed velocities are restricted to a *discrete* set of values. If it is impossible to *explain away* the observed strange feature, then at the end of the day one is left with only one choice: *to accept the observation*. In such circumstances, it would be logically impossible to claim that one is actually *measuring* the velocity of a bullet. One will then only be entitled to *define* the velocity of the bullet in terms of the *action* that was hitherto being called the velocity measurement. Consequently the *act*, using the gun and the target surface, would be *defined* as the "velocity measurement". Of course, this is no longer really a *measurement* but only an act that produces something as an outcome that we *choose to call* the velocity of the bullet[6].

The bottom line is that there is no *logical* reason to call our experiment, "velocity measurement of bullets" but only a *psychological* one arising out of the fact that we started out *believing* that we are measuring the velocity of bullets[7].

Basic Concepts in Quantum Mechanics

We will now lay down the concepts that will constitute the building blocks of QM. If we have been able to convey the spirit of the game in the foregoing sections then it should be easy to see the motivation for the definitions that follow.

The concept of measurement

As we have tried to explain in the previous sections measurement is the most fundamental concept in QM. Everything that follows is built from this key concept. Naturally, we shall start by defining what we mean by "measurement" in QM.

Measurement *By a measurement in QM, we mean a well defined set of actions (usually involving some apparatus) to which we can associate a well defined set of real numbers that are called the outcomes of the measurement.*

---

[6]Here the term *outcome* is also some part of the act of measurement that we *call* the "outcome".

[7]This is nicely summed up in a famous quote by Julian Schwinger: "We only borrow names".

The concept of observable

For reasons that we have already clarified, in QM, we can no longer conceive an "observable" as an *observable property* of a *system* (a concept for which we have no definition as yet) as we are used to doing in classical physics. Observables in QM are actually defined in terms of measurements.

Observable *An observable in QM is characterized by a well defined collection of measurements*[8]. *A collection of such measurements are required to share the same set of outcomes.*

The set of all the outcomes of an observable is called the **spectrum** of the observable[9].

In practice, the measurement prescriptions that define an observable are chosen by the way we *think* we could measure the value of the observable, usually, using some classical reasoning. However, the role of the measurement is not really to *measure* the value of an observable. In QM, it actually *defines* the observable.

The concept of state

The concept of *state* always comes into being in the context of a *theory*. Typically, by a *"state"* we mean some minimal information, pertaining to the *system* under question, from which the theory can extract all the information about the system within the scope of the theory. Thus, to talk about *state,* we need to have some underlying *theory,* and some notion of a *system* at the back of our mind. The theory on which the definition of a *quantum state* is based is founded on the following law:

- *There exists sets of measurement processes such that if one makes two successive measurements from the set (which may or may not correspond to distinct observables), then a definite probability can be assigned to every possible ordered pair of outcomes, provided that the interval between the measurements is held fixed*[10].

We shall call this the **law of definite probabilities**. It will form the basis of our entire discourse. Note that the law states that definite probabilities exist *only for specified sets* of measurement processes. This is because this law *intends* to make a statement about measurements on a given *system*. It is

---

[8]In this definition, we use a *collection* of measurements instead of just one measurement to allow for the fact that just as, classically, there can be more than one experimental setup to *measure* the same observable, in the quantum context, a quantum state (to be defined shortly) can be *prepared* in several ways.

[9]Obviously, I do not intend to mean that every measurement process that has the same set of outcomes must necessarily correspond to the same observable but only that measurements corresponding to the same observable must have the same set of outcomes.

[10]This is not the most general scenario but we shall assume this to be true for now.

only to successive measurements on a given system that we expect the law of definite probabilities to hold. Measurements on different systems are obviously expected to be uncorrelated. But we do *not* yet have a definition for a *system*. So, in absence of a concrete definition of a system, we assert that the law of definite probabilities holds only within certain *sets* of measurements. This means we are actually, implicitly, identifying a system with a well defined set of measurement processes to which the law of definite probabilities apply[11].

In this and the next two chapters we shall restrict ourselves to a special scenario called *"instantaneously subsequent measurements"*[12]. Let me explain what I mean by this. If there is a delay between successive measurements, then, in general, the probabilities will be dependent on the extent of the delay. This means that the probabilities (for pairs of measurement outcomes) are not *uniquely* specified by the outcomes alone. One has to also specify the delay between the measurements. However, it is possible to imagine that the duration between measurements can be made so small that a further reduction in the duration between measurements do not lead to different probabilities. Such sequences of measurements will be called *instantaneously subsequent*. If we restrict ourselves to instantaneously subsequent measurements, a unique probability can be assigned to every ordered pair of measurement outcomes.

The law of definite probabilities allows us to conceive a simple theory that is aimed at predicting the probabilities for instantaneously subsequent measurements. In this simple theory, all we need to do is determine and record, once and for all, the probability for every possible pair of measurement outcomes for every sequence of two instantaneously subsequent measurements. This will enable us to predict the probability of producing a pair of outcomes upon two successive measurements (of arbitrary observables which may or may not be distinct), at all future times. We will call this the *primitive theory*.

We also observe that although we do not as yet have the precise definition of a system, we do have a tentative notion for it. As we have mentioned above, in our mind we can associate a system with the set of measurements to which the law of definite probabilities apply. It is for such sets of measurements that we have our primitive theory.

Now, in the primitive theory, it is the last measurement result that determines the probability of the outcomes of an arbitrary instantaneously subsequent measurement, and this is the sole issue that the theory is concerned with. This naturally leads us to a definition of a quantum state.

---

[11]This is a temporary notion which shall be made precise shortly.

[12]As far as I know, the coinage of this extremely useful jargon is due to Marvin Chester: (Chester 1987).

State *A state in QM is characterized by the last measurement result*[13].

Our definition of a quantum state thus amounts to making the following assertion:

- *A measurement is a state preparation process and every measurement prepares a state that is completely characterized by the outcome of the measurement*[14].

The concept of system

Having defined states as preparations through measurements, we will now provide a more concrete definition of a quantum system in terms of states. Whatever be our intuitive notion of a system, we will all agree that a system must necessarily exist in one of its states. It is impossible to imagine a system that is *not* in one of its states. The universal set of states that a system can inhabit can therefore be used as a characterization of the system. Hence, we move to define a quantum system as follows:

System *A system in QM will be characterized by the entire collection of states that are associated with a set of measurement processes to which the law of definite probabilities apply.*

The concept of event

In probability theory, a *sample space* is defined to be the set of all possible outcomes of an experiment. An *event* is characterized by a subset of the sample space. An *elementary event* is a subset that contains just one element - it is essentially one point in the sample space[15]. In QM, the typical experiment that one considers is a measurement of some observable on a well defined state of a system. The experiment produces some outcome that labels the final state of the system. It is thus natural to characterize the elementary events of these experiments by the *initial* and *final* states. Unless otherwise specified, in our discussion, an event will always mean an elementary event of such an experiment. If we restrict ourselves to instantaneously subsequent measurements, such experiments are essentially defined by a pair of measurements (where the role of the first measurement is to prepare the state). So let us agree to define an (elementary) event in QM as follows:

---

[13]Note that our definition is quite consistent with the way we use the word *"state"* in everyday life. When a doctor asks his nurse about the *state* of his patient, the nurse normally reports the latest *observations* (the last measured blood pressure, temperature, etc.) made on the patient. The most recent observation is believed to be the most relevant.

[14]Actually, a state can always be characterized by the prescription of its preparation. In QM, additionally, we identify measurements with the state preparation process.

[15]Events that are not represented by singleton subsets (i.e., subsets containing just one element) are often called *compound events* to distinguish them from elementary events.

Event *An event in QM is characterized by an ordered pair of states associated with some system*[16].

The initial and final states need not necessarily be distinct. The trivial event when there is no change of state is characterized by identical initial and final states. This is the case when successive measurements of the same observable yields the same outcome.

This is a good point to pause and fix the notation.

Notation

An observable will be denoted by placing a hat on an alphabet: $\hat{\psi}, \hat{\phi}, \hat{A}, \hat{b}$, etc.

The outcomes of measurement of an observable $\hat{\psi}$ will be denoted by ψ_i.

The spectrum of an observable $\hat{\psi}$ will be denoted by $\{\psi_i\}$ where the index 'i' is assumed to run over the entire spectrum of $\hat{\psi}$.

A state characterized by the measurement result ψ_i will be denoted by $|\psi_i\rangle$.

An event characterized by the initial state $|\psi_i\rangle$ and final state $|\phi_j\rangle$ will be denoted by $|\psi_i\rangle \to |\phi_j\rangle$. Often we shall use the shorthand $\psi_i \to \phi_j$ if there is no possibility of confusion.

The probability of the event $\psi_i \to \phi_j$ will be denoted by $P[\psi_i \to \phi_j]$.

Illustration

Let us go back to the system of professor Funnyman's ghosts. We consider three observables for ghosts: colour, odour and taste. We would denote them by

colour: $\hat{C}$,   odour: $\hat{O}$,   taste: $\hat{T}$.

---

[16] For more general scenarios an event can still be characterized by initial and final states but state itself is defined more generally. We shall furnish this definition later.

Let the spectra of the respective observables be[17]

$\{C_1, C_2\} \equiv \{\text{red, green}\}, \quad \{O_1, O_2\} \equiv \{\text{coffee, lemon}\}, \quad \{T_1, T_2\} \equiv \{\text{bitter, sweet}\}.$

Examples of states of the system would then be

$|C_1\rangle \equiv |\text{red}\rangle, \quad |O_2\rangle \equiv |\text{lemon}\rangle, \quad |T_2\rangle \equiv |\text{sweet}\rangle, \dots, \text{etc.}$

We will often refer a ghost-state such as $|\text{red}\rangle$ as a "red ghost".

Typical events for the ghost would be written as

red $\rightarrow$ sweet, lemon $\rightarrow$ bitter, bitter $\rightarrow$ bitter, $\dots$, etc.,

with the respective probabilities denoted by

$P[\text{red} \rightarrow \text{sweet}], \quad P[\text{lemon} \rightarrow \text{bitter}], \quad P[\text{bitter} \rightarrow \text{bitter}], \dots, \text{etc.}$

Now that we have constructed the basic definitions and notations, let us, once more, state the central questions in QM using the vocabulary and notation we have developed (see Figure 2.5).

If we make a measurement of an arbitrary observable $\hat{\phi}$ on a state $|\psi_i\rangle$ what are the possibilities $\{\phi_j\}$ in the outcome of the measurement (i.e., what is the spectrum of $\hat{\phi}$) and what are the probabilities $P[\psi_i \rightarrow \phi_j]$ of the different possibilities ϕ_j?[18]

Laws of Quantum Mechanics

The *measure of goodness* of a theory is *how much can be embodied in how little*. This is called the *predictive power* of the theory. Clearly, our primitive theory based on the law of definite probabilities has a very low predictive power. This theory requires far too many inputs than we can hope to tolerate.

---

[17]We must keep in mind that in an actual situation, outcomes of measurements will always be real numbers.

[18]For instantaneously subsequent measurements $\hat{\phi}$ is assumed to be measured soon after the state $|\psi_i\rangle$ is prepared. However, in the general case the state $|\psi_i\rangle$ may be prepared at some arbitrary earlier time.

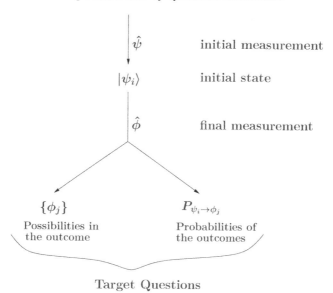

Target Questions

FIGURE 2.5: The typical setting of a quantum mechanical problem and the "Target Questions".

If there are M observables each having N elements in their spectra, then we need to specify $M \times N$ possible outcomes and $(M \times N)^2$ probabilities. The primitive theory can, however, serve as a reference point to assess how much progress we have actually made when we try to build a more *improved* theory. In this section we shall introduce more laws of QM that will lead to such an improved theory - a theory that will require fewer inputs than the primitive theory. To introduce these new laws, we shall first need to recast the law of definite probabilities in a slightly different way.

Law of Definite Probabilities *To every event $\psi_i \to \phi_j$ one can associate a complex number $\langle \phi_j | \psi_i \rangle$ such that the probability $P[\psi_i \to \phi_j]$ of the event is given by*

$$P[\psi_i \to \phi_j] = |\langle \phi_j | \psi_i \rangle|^2$$

*The complex number $\langle \phi_j | \psi_i \rangle$ is called the **probability amplitude** (or amplitude in short) of the event $\psi_i \to \phi_j$.*

Since the *total probability* must be unity, the probability amplitudes must satisfy what we will call the **totality condition**:

$$\sum_k |\langle \chi_k | \psi_i \rangle|^2 = 1$$

if k runs over the *entire* spectrum of the observable $\hat{\chi}$.

In the literature, this law is referred to as the **Born rule,** after German physicist Max Born, who first correctly put forward the probabilistic interpretation of quantum mechanics.

Illustration

The probability of a red ghost to be found in the $|\text{sweet}\rangle$ state upon measurement of taste will be given by

$$P\left[\text{red} \rightarrow \text{sweet}\right] = |\langle \text{sweet}|\text{red}\rangle|^2$$

where $\langle \text{sweet}|\text{red}\rangle$ is the probability amplitude for the event $|\text{red}\rangle \rightarrow |\text{sweet}\rangle$. The totality conditions applied to the $|\text{red}\rangle$ state would read

$$|\langle \text{bitter}|\text{red}\rangle|^2 + |\langle \text{sweet}|\text{red}\rangle|^2 = 1$$
$$|\langle \text{coffee}|\text{red}\rangle|^2 + |\langle \text{lemon}|\text{red}\rangle|^2 = 1$$
$$|\langle \text{red}|\text{red}\rangle|^2 + |\langle \text{green}|\text{red}\rangle|^2 = 1$$

Note that at this point the above statement of the law of definite probabilities is only a cumbersome way of saying that to every event we can assign a definite probability. Of course, in order to make this assertion there was no need to assign a complex number to every event. A real number in the interval $[0, 1]$ would have sufficed. The actual reason for doing this is to enable us to describe a weird fact about quantum probabilities that is responsible for almost all of the mystery that QM has to offer. We state this fact as our next law.

Superposition Principle *Probability amplitudes satisfy the relation*

$$\langle \phi_j|\psi_i\rangle = \sum_k \langle \phi_j|\chi_k\rangle \langle \chi_k|\psi_i\rangle$$

where k runs over the entire spectrum of any arbitrary observable $\hat{\chi}$.

The superposition principle is a relationship between certain probability amplitudes. It is not difficult to see that it implies a relationship between probabilities that resemble the relationship between intensities of interfering waves. This has led to the common jargon: *"quantum probabilities interfere".* Let me demonstrate this more clearly. Observe that the right hand side of the above formula is a sum of products of amplitudes $\langle \phi_j|\chi_k\rangle \langle \chi_k|\psi_i\rangle$. Now the probability of the composite event, $\psi_i \rightarrow \chi_k$ followed by $\chi_k \rightarrow \phi_j$, is given by

$$P\left[\psi_i \rightarrow \chi_k\right] P\left[\chi_k \rightarrow \phi_j\right] = |\langle \phi_j|\chi_k\rangle|^2 |\langle \chi_k|\psi_i\rangle|^2 = |\langle \phi_j|\chi_k\rangle \langle \chi_k|\psi_i\rangle|^2$$

Thus the product $\langle\phi_j|\chi_k\rangle\,\langle\chi_k|\psi_i\rangle$ can be considered to be the amplitude for the composite event $\psi_i \to \chi_k$ followed by $\chi_k \to \phi_j$. Such a composite event can be visualized as a *path*: $\psi_i \to \chi_k \to \phi_j$. The right hand side of the superposition principle then becomes a sum of amplitudes for all possible paths $\psi_i \to \chi_k \to \phi_j$ (each path being characterized by a distinct intermediate state labelled by χ_k; see Figure 2.6). Now recall that in the theory of interfering waves, to get the resultant intensity of waves arriving at some point through different paths, we add the displacements for each path and then take the square of the modulus of this sum[19]. The parallel is evident. *Quantum probabilities combine like the intensities of interfering waves: The probability amplitudes are to probabilities what displacements are to intensities*[20].

We wish to emphasize that *it is only if we choose **not** to make the intermediate measurement that the superposition principle will be applicable.* If we do make the intermediate measurement the probability is given by the classical probability addition rule. This means that if we make three successive measurements of the observables $\hat\psi$, $\hat\chi$ and $\hat\phi$, the probability $P\,[\psi_i \to \{\chi_k\} \to \phi_j]$ that the outcome of the first measurement is ψ_i and that of last is ϕ_j (irrespective of the outcome χ_k in the intermediate measurement) is given by

$$P\,[\psi_i \to \{\chi_k\} \to \phi_j] = \sum_k |\langle\phi_j|\chi_k\rangle\,\langle\chi_k|\psi_i\rangle|^2$$

Here, and in the rest of the chapter k (the subscript of χ) will be assumed to run over the *entire* spectrum of $\hat\chi$ in summations.

---

[19]The square of the *modulus* is actually more appropriate when the displacements are taken to be complex as is sometimes done in the theory of interfering electromagnetic (light) waves. In this case the parallel that we are trying to point out is even closer.

[20]For the sake of visualization (see Figure 2.6), it may be useful to imagine the observables $\hat\psi$, $\hat\chi$ and $\hat\phi$ to be associated with a measurement of position of particles (let us *call* them *photons*) along the y-direction (along the slits) at three different x-positions (perpendicular to the slits), say x_1, x_2 and x_3. So we are interested in the event that photons are ejected from a certain y-position y_i at x_1 and detected at a certain y-position y_j at x_3. We can imagine, at an intermediate position x_2, a series of slits with sensors placed at edge of each slit which fire when the particle passes through the slit. This corresponds to the $\hat\chi$ measurement. It turns out that for some reason, the only positions through which the particles may pass are the locations of the slits $y_1, y_2, \ldots, y_n$. One may argue that this is an artifact of the sieve-like y-measurement apparatus (which allows photons to pass through particular positions only) and *not* a property of the photons. The remarkable observation is, however, that irrespective of the design of the $\hat y$-measurement apparatus, the allowed y-positions at x_2 are always the same if we intend to detect all possibilities. The parallel can be carried further by noting that the intensity of a stream of photons (if each photon is assumed to carry some definite energy), at the point ϕ_j on the detector $\hat\phi$, is expected to be proportional to the number of photons reaching that point per second, and this number would, in turn, be proportional to the probability of a single photon reaching that point of the detector. Now, it just takes one leap of the imagination to arrive at the superposition principle: the observables $\hat y\,(x_1), \hat y\,(x_2)$ and $\hat y\,(x_3)$ need to be generalized to arbitrary observables $\hat\psi$, $\hat\chi$ and $\hat\phi$. For a delightful exposition of the interference phenomena as a motivation for QM, see (Feynman et al. 1963); Indeed, much of the present work is inspired by this very celebrated course of lectures by Richard P. Feynman.

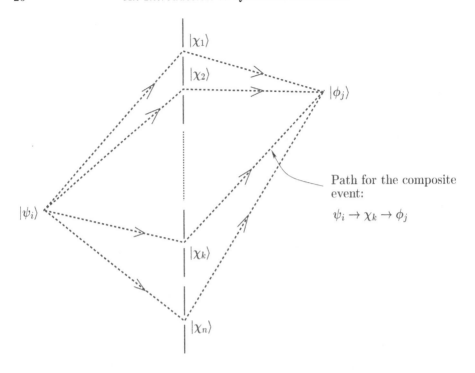

FIGURE 2.6: Superposition Principle: $\langle \phi_j | \psi_i \rangle = \sum_k \langle \phi_j | \chi_k \rangle \langle \chi_k | \psi_i \rangle$.

The nontrivial feature of the superposition principle is to replace the sum of squares by the square of the sum when the intermediate measurement is not performed:

$$P[\psi_i \to \phi_j] = \left| \sum_k \langle \phi_j | \chi_k \rangle \langle \chi_k | \psi_i \rangle \right|^2$$

It is often remarked that the superposition principle is a violation of the classical probability addition rule and that it is here that quantum mechanics actually departs from classical physics. In our opinion this is incorrect. As we have explained above, the classical probability addition formula and the superposition principle apply to two different situations. The left hand side of the last formula refers to an elementary event $\psi_i \to \phi_j$ in an experiment involving *two* successive measurements ($\hat{\psi}$ and $\hat{\phi}$) while the one preceding that refers to an event $\psi_i \to \{\chi_k\} \to \phi_j$ in an experiment involving *three* successive measurements ($\hat{\psi}$, $\hat{\chi}$ and $\hat{\phi}$). The event $\psi_i \to \{\chi_k\} \to \phi_j$ is a collection of mutually exclusive, simpler events where each simple event is of the form $\psi_i \to \chi_k$ followed by $\chi_k \to \phi_j$, and the collection is generated by making k run over the entire spectrum of $\hat{\chi}$. It is to this collection of mutually exclusive events that the classical probability addition rule should apply, as it rightfully does in this case. The *composite events* $\psi_i \to \chi_k$ followed by $\chi_k \to \phi_j$ are not mutually exclusive ways by which the event $\psi_i \to \phi_j$ is realized. They are mutually exclusive ways by which the event $\psi_i \to \{\chi_k\} \to \phi_j$ is realized[21].

Illustration

If we apply the superposition principle to the amplitude $\langle \text{lemon}|\text{red}\rangle$ for our ghost system, we get

$$\langle \text{lemon}|\text{red}\rangle \;=\; \langle \text{lemon}|\text{bitter}\rangle \langle \text{bitter}|\text{red}\rangle$$
$$+ \langle \text{lemon}|\text{sweet}\rangle \langle \text{sweet}|\text{red}\rangle$$

where we have used the taste states as intermediate states.

Law of Sustained States *Instantaneously subsequent measurement of the same observable yields the same outcome.*

$$P\left[\psi_i \to \psi_i\right] = 1$$

The law of sustained states trivially translates into the following statement in terms of amplitudes:

$$\langle \psi_j|\psi_i\rangle = \delta_{ij}$$

where ψ_i and ψ_j are two outcomes of the same observable $\hat{\psi}$ and δ_{ij} is the Kronecker delta function[22].

One might think that one should more generally assume $\langle \psi_j|\psi_i\rangle = \delta_{ij}e^{i\theta}$. It will soon be clear that one can, without any loss of generality choose the phase $\theta = 0$.

---

[21]We are not taking anything away from the importance of the superposition principle. It is certainly true that nothing like the superposition principle exists in classical probability theory. We just wish to underline that the context of its applicability does not imply a violation of the classical probability addition rule.

[22]For those unfamiliar, the kronecker delta function is defined as follows:

$$\delta_{ij} = \begin{cases} 0 & \text{for } i \neq j \\ 1 & \text{for } i = j \end{cases}$$

Illustration

For the ghosts, the law of sustained states mean, for example, that if we measure colour on a red ghost the outcome is certain to be red. For the colour states this implies

$$\langle green|red\rangle = 0$$
$$\langle red|red\rangle = 1$$
$$\langle green|green\rangle = 1$$
$$\langle red|green\rangle = 0$$

Exchange Symmetry   *Probability amplitude of events, $\psi_i \to \phi_j$ and $\phi_j \to \psi_i$, with the initial and final states exchanged are complex conjugates of each other.*

$$\langle\phi_j|\psi_i\rangle^* = \langle\psi_i|\phi_j\rangle$$

The most obvious implication of this law is that the probability of events with initial and final states swapped are equal:

$$P[\psi_i \to \phi_j] = P[\phi_j \to \psi_i]$$

Though this consequence is easy to see, it is not apparent what suggests the complex conjugation. We wish to caution the reader that the amplitudes are *not* equal under exchange of initial and final states, only the probabilities are. This has nontrivial and measurable consequences[23].

To see what prompts the complex conjugation, note that the totality condition can be written as

$$\sum_k \langle\chi_k|\psi_i\rangle^* \langle\chi_k|\psi_i\rangle = 1$$

Again, the superposition principle and the law of sustained states, $\langle\psi_i|\psi_i\rangle = 1$, taken together gives

$$\sum_k \langle\psi_i|\chi_k\rangle \langle\chi_k|\psi_i\rangle = 1$$

---

[23]Check that the superposition principle leads to distinct probabilities if we drop the complex conjugation and use $\langle\phi_j|\psi_i\rangle = \langle\psi_i|\phi_j\rangle$ instead.

The simplest way to render these equations consistent is to require

$$\langle\chi_k|\psi_i\rangle^* \;=\; \langle\psi_i|\chi_k\rangle$$

Since there is nothing special about the states $|\psi_i\rangle$ or $|\chi_k\rangle$, similar results are expected to hold for arbitrary pairs of states. Of course, mathematical consistency is only a *necessary condition* for a physical law and cannot by itself validate the law. Such validation can only come from experiments. Indeed, exchange symmetry has been validated by experiments.

Illustration

For our ghost system, exchange symmetry would imply, for example, that the probability of getting a sweet ghost upon measurement of taste on a red ghost is the same as that of getting the outcome red on measurement of colour on a sweet ghost. The amplitudes for the two events are however conjugates of each other:

$$\langle\text{sweet}|\text{red}\rangle \;=\; \langle\text{red}|\text{sweet}\rangle^*$$

Finally, before closing this section let us note that the totality condition, superposition principle and exchange symmetry naturally leads to the law of sustained states:

$$\sum_k |\langle\chi_k|\psi_i\rangle|^2 \;=\; 1$$

$$\sum_k \langle\chi_k|\psi_i\rangle^* \langle\chi_k|\psi_i\rangle \;=\; 1$$

$$\sum_k \langle\psi_i|\chi_k\rangle \langle\chi_k|\psi_i\rangle \;=\; 1$$

$$\langle\psi_i|\psi_i\rangle \;=\; 1$$

Thus, the law of sustained states is rendered redundant. It becomes a *theorem* that follows from the other three laws of QM. What then was the point of introduction of the law of sustained states? Well, it was to motivate the exchange symmetry law which is not very obvious by itself. But we have seen in the preceding paragraph, how a requirement of consistency of the totality condition, superposition principle and law of sustained states leads us naturally to assume the existence of exchange symmetry. However, after having assumed exchange symmetry, it now makes no sense to retain the status of sustained states as a *law*. Henceforth, we shall only refer to the other three laws as the independent laws of QM.

The Quantum State Redefined

The laws of QM discussed in the previous section imply that for instantaneously subsequent measurements, if for all states $|\psi\rangle$ we furnish the set of probability amplitudes $\{\langle\chi_k|\psi\rangle\}$ with k running over the entire spectrum of an *arbitrary but specific* observable $\hat{\chi}$, the probability $P[\alpha_i \to \beta_j]$ of an arbitrary event $\alpha_i \to \beta_j$ can be computed in a straight forward manner:

$$
\begin{aligned}
P[\alpha_i \to \beta_j] &= |\langle\beta_j|\alpha_i\rangle|^2 \\
&= \left|\sum_k \langle\beta_j|\chi_k\rangle\langle\chi_k|\alpha_i\rangle\right|^2 \\
&= \left|\sum_k \langle\chi_k|\beta_j\rangle^*\langle\chi_k|\alpha_i\rangle\right|^2
\end{aligned}
$$

where we have made use of the superposition principle and exchange symmetry in the second and third lines respectively.

Thus, the laws of QM have led us to an *improved theory* (in comparison to our primitive theory). To predict the probabilities, it is now enough to specify the probability amplitudes $\langle\chi_k|\psi\rangle$ only for those events $\psi_i \to \chi_k$ that has $|\chi_k\rangle$ as a final state. Comparing this theory with the primitive theory (that was based on the law of definite probabilities alone) we see that for this theory the required number of inputs is greatly reduced. Once again, if there are M observables each having N elements in their spectra, we now need to specify $M \times N^2$ amplitudes in order to determine all possible amplitudes. This theory is therefore, surely, an improvement over the primitive theory which required $(M \times N)^2$ inputs[24].

Definition of state in the improved theory

As we have mentioned before, in any physical theory the *state* of a system usually refers to the minimal set of information pertaining to the system that enables the theory to extract the maximal information about the system within the scope of the theory. The development outlined in the preceding paragraph thus clearly leads to the following definition of a state:

- *A state $|\psi\rangle$ in the χ space is defined to be the ordered list of probability amplitudes $\{\langle\chi_k|\psi\rangle\}$, where k runs over the entire spectrum of an arbitrary but specific observable $\hat{\chi}$.*

---

[24]If we take the law of sustained states into account, the improvement is even better. But we are not interested in that arithmetic here. It is enough if we are convinced that the improvement is overwhelming.

Note that, since this way of defining a state makes reference to an (arbitrary) observable $\hat{\chi}$, we refer to the state as the *"state in the χ space"*.

I wish to emphasize that having defined states as a list of probability amplitudes, our earlier characterization of a state using the last measurement result now merely becomes a way of *labeling* a state. Moreover, we no longer need to restrict states to be prepared only by measurement processes. In fact, our new definition of a quantum state implies that any process that ensures that subsequent measurements are in accordance with a list of probability amplitudes, say $\{\langle\chi_k|\psi\rangle\}$ with k running over the entire spectrum of some observable $\hat{\chi}$, is a legitimate state preparation process. More explicitly, this means that we are now identifying the state with a condition that is characterized by the set of complex numbers $\{\langle\chi_k|\psi\rangle\}$ such that upon a measurement of an arbitrary observable $\hat{\phi}$ the probability of getting an outcome ϕ_j will be given by $\left|\sum_k \langle\chi_k|\phi_j\rangle^* \langle\chi_k|\psi\rangle\right|^2$. Incidentally, one such process is *"waiting"* where one merely waits and lets one quantum state evolve in time into a new quantum state. We shall take this up in a later chapter. Of course, we can always ask whether to every possible quantum state there corresponds an observable such that upon its measurement it may prepare the system in the given state. We shall see that, in principle, this is true. To see how this comes about, however, we have to wait until we are done with the next chapter[25].

Illustration

The ghost-states now become ordered pairs. For example, the ghost state $|\text{coffee}\rangle$ in the colour space will be the ordered pair,

$$\{\langle\text{red}|\text{coffee}\rangle, \langle\text{green}|\text{coffee}\rangle\}$$

Incidentally, if we express $|\text{coffee}\rangle$ in the odour space the ordered pair is

$$\{\langle\text{coffee}|\text{coffee}\rangle, \langle\text{lemon}|\text{coffee}\rangle\} = \{1, 0\}$$

by virtue of the law of sustained states. Now it is clear that if we are interested in the probability of the event $|\text{coffee}\rangle \rightarrow |\text{sweet}\rangle$, it is adequate to specify the states $|\text{coffee}\rangle$ and $|\text{sweet}\rangle$ in some specified space, say colour. Then the prescription outlined above gives

$$P[\text{coffee} \rightarrow \text{sweet}] = |\langle\text{sweet}|\text{coffee}\rangle|^2$$

---

[25]The careful reader will have noticed that unless we assume that the size of spectra of all observables are the same, states in different spaces will have different number of elements and our construction will break down. Obviously, it would be extremely unnatural to assume that all observables for a system will have exactly the same number of possible outcomes. Unfortunately, we will be able to resolve this issue only in the fifth chapter. Until then, we will have to stick to the unnatural scenario that size of the spectra of all observables are equal.

where the amplitude $\langle\text{sweet}|\text{coffee}\rangle$ is given by

$$
\begin{aligned}
\langle\text{sweet}|\text{coffee}\rangle \;&=\; \langle\text{sweet}|\text{red}\rangle\,\langle\text{red}|\text{coffee}\rangle \\
&\qquad +\, \langle\text{sweet}|\text{green}\rangle\,\langle\text{green}|\text{coffee}\rangle \\
&=\; \langle\text{red}|\text{sweet}\rangle^{*}\,\langle\text{red}|\text{coffee}\rangle \\
&\qquad +\, \langle\text{green}|\text{sweet}\rangle^{*}\,\langle\text{green}|\text{coffee}\rangle
\end{aligned}
$$

There is a neat way to express the last equation using matrices:

$$
\langle\text{sweet}|\text{coffee}\rangle \;=\; \begin{bmatrix} \langle\text{red}|\text{sweet}\rangle \\ \langle\text{green}|\text{sweet}\rangle \end{bmatrix}^{\dagger} \begin{bmatrix} \langle\text{red}|\text{coffee}\rangle \\ \langle\text{green}|\text{coffee}\rangle \end{bmatrix}
$$

where the states have been expressed as column matrices and the symbol '$\dagger$' stands for Hermitian conjugation.

Chapter 3

States as Vectors

Our description of QM will now go through several stages of metamorphosis. We shall rephrase QM in a language that will increasingly become more sophisticated (i.e., mathematical) and obscure (i.e., connection of statements to the facts they describe will become more involved). In the last chapter we introduced QM in an *empirical* language. It was done to make the factual content of QM apparent. Now we shall try to find a mathematical characterization of all the concepts and facts that were introduced in the last chapter[1]. This will lead us to the *formal* description of QM. The point of introducing QM in an empirical language was to avoid introducing QM through the formal language from the outset, so that we can demonstrate how the empirical content can be translated into the formal language and we may be able to appreciate the power and beauty of what this translation accomplishes. It will also allow us to identify which parts of the formal description of QM are indispensable (i.e., required by nature) and which are required by our choice of the language. We shall start by learning some linear algebra.

Vectors in $\mathbb{C}^n$

Consider the set $\mathbb{C}^n$ comprising all n-tuples of complex numbers[2] One can choose to write the elements of $\mathbb{C}^n$ as n-dimensional complex column matrices. We shall call these objects, **vectors**. We will denote vectors by alphabets in square brackets (e.g., $[A]$, $[B]$,..., etc.) and their components by placing a superscript on the alphabet (e.g., A^i, B^i,..., etc.). Complex numbers will be referred to as **scalars** and denoted by lower case alphabets (e.g., a, b,..., etc.[3]).

---

[1] Probably the earliest and the most authentic work in this direction is due to P.A.M. Dirac; see (Dirac 1958).

[2] A n-tuple is a set of n elements (in this case complex numbers) arranged in a definite order.

[3] Everything that we will do in this section can also be done with n-tuples of real numbers which would belong to the set $\mathbb{R}^n$.

Addition of vectors and multiplication of a vector by a scalar

We define **addition of vectors** in $\mathbb{C}^n$ and **multiplication of a vector by a scalar** (sometimes called *scalar multiplication* in short) by the usual rule of matrix addition and multiplication of a matrix by a scalar:

$$[A] + [B] = \begin{bmatrix} A^1 + B^1 \\ A^2 + B^2 \\ \cdots \\ \cdots \\ A^n + B^n \end{bmatrix} \qquad c\,[A] = \begin{bmatrix} cA^1 \\ cA^2 \\ \cdot\cdot \\ \cdot\cdot \\ cA^n \end{bmatrix}$$

Addition of vectors and multiplication of a vector by a scalar have the following algebraic properties:

1. Associativity of addition of vectors: For all vectors $[A]$, $[B]$, $[C]$

$$([A] + [B]) + [C] = [A] + ([B] + [C])$$

2. Existence of Additive Identity: There exists a vector $[0]$, that is called the **additive identity**, such that for any vector $[A]$

$$[A] + [0] = [0] + [A] = [A]$$

3. Existence of Additive Inverse: For every vector $[A]$ there exists a vector $[-A]$, that is called the **additive inverse** of $[A]$, such that

$$[A] + [-A] = [-A] + [A] = [0]$$

4. Commutativity of addition of vectors: For all vectors $[A]$, $[B]$

$$[A] + [B] = [B] + [A]$$

5. Compatibility of multiplication of scalars with multiplication of a vector by a scalar: For every vector $[A]$ and all scalars a, b

$$a\,(b\,[A]) = (ab)\,[A]$$

6. Distributivity of scalar multiplication of a vector over addition of vectors: For all vectors $[A]$ and $[B]$ and every scalar a

$$a\,([A] + [B]) = a\,[A] + a\,[B]$$

7. Distributivity of scalar multiplication of a vector over addition of scalars: For every vector $[A]$ and all scalars a, b

$$(a + b)\,[A] = a\,[A] + b\,[A]$$

8. Equality of identity of scalar multiplication and multiplicative identity of scalars: For every vector $[A]$

$$1\,[A] = [A]$$

Inner product of vectors

We define an ***inner product*** (also called *scalar product*) of two vectors $[A]$ and $[B]$ denoted by $([A], [B])$ (or more often, simply, as (A, B)) in the following way:

$$(A, B) \;=\; [A]^{\dagger} [B]$$

It is easy to check that this inner product has the following properties[4]

1. Conjugate Symmetry: For all vectors $[A]$, $[B]$

$$(A, B) = (B, A)^{*}$$

2. Linearity: For all vectors $[A]$, $[B]$, $[C]$ and all scalars b, c

$$([A], b[B] + c[C]) = b(A, B) + c(A, C)$$

3. Positive-definiteness: For any vector $[A]$

$$(A, A) \geq 0 \quad \text{and} \quad (A, A) = 0 \quad \text{if and only if} \quad [A] = [0]$$

Norm

We define the ***norm*** $|A|$ of a vector $[A]$ by the positive square root of the scalar product of the vector with itself:

$$|A| \;=\; \sqrt{(A, A)}$$

*A vector that has unit norm is said to be **normalized**.* Vectors that are not normalized can be multiplied by appropriate scalars to give normalized vectors. This process is called *normalization*: Suppose we have a vector $[A]$ such that

$$|A| \;=\; N$$

where N is some real number. We can always construct the new vector

$$[A_N] = \frac{1}{N}[A]$$

which is normalized:

$$|A_N| \;=\; 1$$

---

[4]It is possible to define a *product* in other ways which will obey the same properties. Later on, we shall generalize the definition and call all such products an *inner product*.

Orthogonality

If a pair of vectors $[A]$ and $[B]$ have a vanishing scalar product:

$$(A, B) = 0$$

then the vectors $[A]$ and $[B]$ are said to be ***orthogonal*** to each other. If the members of a set of vectors $A = \{[A_1], [A_2], \ldots, [A_n]\}$ are mutually orthogonal:

$$(A_i, A_j) = 0 \qquad \text{when} \quad i \neq j$$

then the set A is said to be an ***orthogonal set***. If the members of a set of normalized vectors $A = \{[A_1], [A_2], \ldots, [A_n]\}$ are mutually orthogonal:

$$(A_i, A_j) = \delta_{ij}$$

then the set A is said to be an ***orthonormal set***.

Basis and dimension

A ***linear combination*** of a set of vectors is a *sum of scalar multiples* of the vectors. For example, if for a set of r scalars c_i and vectors $[B_i]$, the vector

$$[A] = \sum_{i=1}^{r} c_i [B_i]$$

then $[A]$ is said to be a linear combination of the vectors $[B_i]$.

Now, consider the set of vectors $E = \{[E_k] ; k = 1, 2, \ldots, n\}$, with the components E_k^i of the vector $[E_k]$ being given by

$$E_k^i = \delta_{ik}$$

Thus, E consists of the vectors

$$[E_1] = \begin{bmatrix} 1 \\ 0 \\ \cdots \\ \cdots \\ 0 \end{bmatrix}, \qquad [E_2] = \begin{bmatrix} 0 \\ 1 \\ \cdots \\ \cdots \\ 0 \end{bmatrix}, \ldots\ldots\ldots\ldots\ldots, [E_n] = \begin{bmatrix} 0 \\ 0 \\ \cdots \\ \cdots \\ 1 \end{bmatrix}$$

It is clear that any vector $[A]$ in $\mathbb{C}^n$ can be written as

$$\begin{bmatrix} A^1 \\ A^2 \\ .. \\ .. \\ A^n \end{bmatrix} = A^1 \begin{bmatrix} 1 \\ 0 \\ \cdots \\ \cdots \\ 0 \end{bmatrix} + A^2 \begin{bmatrix} 0 \\ 1 \\ \cdots \\ \cdots \\ 0 \end{bmatrix} + \cdots\cdots + A^n \begin{bmatrix} 0 \\ 0 \\ \cdots \\ \cdots \\ 1 \end{bmatrix}$$

Thus, in other words, any vector $[A]$ in $\mathbb{C}^n$ can be expressed as a linear combination of the vectors in E. We express this by saying that the set of vectors E is a *complete set*. *If an arbitrary vector can be expressed as a linear combination of the vectors in some subset, we say that the subset forms a* **complete set**. Conversely, the set of all linear combinations of some subset is called the **linear span** of the subset.

We observe further, that it is impossible to express any of the vectors in the set E as a linear combination of the other vectors in the set. This is expressed by saying that E is a *linearly independent set*. Thus *if it is impossible to express the vectors in some set as a linear combination of other vectors in the set, we say that the set is a* **linearly independent set**.

Now we come to the most important definition of this subsection. *A set which is complete and linearly independent is called a* **basis**. In general many (actually infinite) such bases exist in $\mathbb{C}^n$. If a basis forms an orthonormal set it is called an *orthonormal basis*. It is evident that the set E comprises an orthonormal basis. The basis E is called the *standard basis* in $\mathbb{C}^n$.

It turns out that every basis in $\mathbb{C}^n$ has exactly n elements and this number is called the **dimension** of $\mathbb{C}^n$.

Problems[5]

1. Consider the set of vectors $S = \left\{ (1,1,3)^T, (1,-1,2)^T, (2,0,3)^T \right\}$ in $\mathbb{C}^3$. Using Cramer's rule express the vector $[\psi] = (1,2,3)^T$ of $\mathbb{C}^3$ as a linear combination of the members of the set S.

2. Show that the set of vectors $\left\{ (1,2,2)^T, (2,1,2)^T, (2,2,1)^T \right\}$ forms a linearly independent set in $\mathbb{C}^3$. Does it form a basis? What about the set $\left\{ (1,2,2)^T, (2,1,2)^T \right\}$?

3. Prove that every basis of $\mathbb{C}^n$ must have exactly n elements.

   (Hint: Prove that the necessary and sufficient condition for an arbitrary vector in $\mathbb{C}^n$ to be expressible as a linear combination of a set of n vectors is that the n vectors must be linearly independent. *This also proves that any set of n linearly independent vectors forms basis in $\mathbb{C}^n$.* Therefore, bases with cardinality larger than n are impossible. Now

---

[5]Throughout the book, in the problems, to save space we have often written the column matrices as transposed row matrices (written with a superscripts T, as usual). Of course, mathematically, row and column matrix are both ordered n-tuples, and one could have used row matrices in place of column matrices all along. It is only a matter of convention.

argue that it is impossible, in general, to express an arbitrary vector in $\mathbb{C}^n$ as a linear combination of less than n vectors[6].)

4. Show that an equivalent definition of the linear independence of a set of vectors $\{[B_1], [B_2], \ldots, [B_r]\}$ is given by the condition:

$$\sum_{i=1}^{r} c_i [B_i] = 0 \quad \Longrightarrow \quad c_i = 0 \quad \text{for all } i$$

where c_i are complex numbers[7].

5. Prove that an orthogonal set is necessarily linearly independent.

Towards a Formal Description

Now we have the preliminary mathematical background to start developing the mathematical characterization of the concepts and features of QM that were introduced in the last chapter.

Description of states

In the last chapter, we started out by characterizing a quantum state by the last measurement result. In the last section of the chapter we saw that a quantum state $|\psi\rangle$ can be characterized by a set of n complex numbers $\{\psi^k; k = 1, 2, \ldots, n\}$ *if* these complex numbers can be interpreted as probability amplitudes: $\psi^k = \langle \chi_k | \psi \rangle$ where χ_k are the members of the spectrum (assumed to consist of n elements) of some arbitrary observable $\hat{\chi}$ of our choice. Since, there could be many such characterizations depending on the choice of the observable $\hat{\chi}$, we called this description, a state in the χ space. Since the amplitudes $\langle \chi_k | \psi \rangle$ that go into the characterization of the state $|\psi\rangle$ must obey the totality condition:

$$\sum_{k=1}^{n} |\langle \chi_k | \psi \rangle|^2 = 1$$

the set of n complex numbers $\{\psi^k; k = 1, 2, \ldots, n\}$ characterizing a state must respect the algebraic condition

$$\sum_{k=1}^{n} |\psi^k|^2 = 1$$

---

[6]You have essentially proved that a basis in $\mathbb{C}^n$ is the smallest complete set and largest linearly independent set.

[7]While the earlier definition, that was provided in the text, is more intuitive, this one comes in much more handy to work with in proofs and computations.

We shall start the construction of the formal description of QM by asserting that *any* set of n complex numbers $\{\psi^k;\ k = 1, 2, \dots, n\}$ for which the above algebraic condition holds may be interpreted as a set of n probability amplitudes $\{\langle \chi_k|\psi\rangle\,;\ k = 1, 2, \dots, n\}$ (with any predetermined choice of $\hat{\chi}$) and therefore can specify a quantum state in the χ space. So, we define a quantum state as follows.

A state in the χ space is defined to be any ordered list of n complex numbers $\{\psi^k; k = 1, 2, \dots, n\}$ which obey the algebraic condition $\sum_{k=1}^{n} |\psi^k|^2 = 1$. The complex numbers ψ^k are interpreted as the probability amplitudes $\langle \chi_k|\psi\rangle$ where k runs over the entire spectrum of the observable $\hat{\chi}$.

It is convenient to write such states as complex column matrices

$$[\psi]^\chi \equiv \begin{bmatrix} \psi^1 \\ \psi^2 \\ .. \\ .. \\ \psi^n \end{bmatrix}$$

where we use the explicit superscript χ in $[\psi]^\chi$ to denote the space that we use for our description. Using the vocabulary introduced in the last section, and noting that the totality condition $\sum_{k=1}^{n} |\psi^k|^2 = 1$ is the same as the normalization condition $([\psi]^\chi)^\dagger [\psi]^\chi = 1$, we can rephrase the above definition of state as follows:

- *A state (in the χ space) is a normalized vector belonging to $\mathbb{C}^n$.*

We shall often refer to a normalized vector in $\mathbb{C}^n$ as a *state vector*. There is one thing we would like to point out here. A normalized vector in $\mathbb{C}^n$ is not really unique. It is defined up to an unimodular scalar factor[8]. This *overall phase* factor is sometimes called a *global phase*. It will be easy to convince ourselves that physical predictions will not depend on *global* phase factors. We emphasize the use of the word "*global*". If we multiply different components of a normalized vector with *distinct* unimodular factors (sometimes called *relative phases*) the norm will still be one, but this will be a physically different state vector[9].

Incidentally, the states $[\chi_k]^\chi$ corresponding to the elements χ_k of the spectrum of $\hat{\chi}$ are simply the members of the standard basis in $\mathbb{C}^n$. This is because, the law of sustained states requires that the components χ_k^i of $[\chi_k]^\chi$ be given by

$$\chi_k^i = \langle \chi_i|\chi_k\rangle = \delta_{ik}$$

---

[8] If $[\psi]$ is a normalized vector (where $\left|\sqrt{(\psi, \psi)}\right| = 1$) then so is $e^{i\theta} [\psi]$ where θ is any real number.

[9] It should now be clear, why we could drop the phase factor $e^{i\theta}$ when we stated the law of sustained states as $\langle \chi_i|\chi_k\rangle = \delta_{ik}$ (instead of $\langle \chi_i|\chi_k\rangle = \delta_{ik} e^{i\theta}$).

In this section, since we shall work exclusively in the χ space, we shall often drop the superscript χ on vectors.

Description of amplitudes

Having constructed the definition of a quantum state, it now becomes our obligation to prescribe how the amplitudes involving such states should be defined in a consistent way. To do this we recall that, by the superposition principle and exchange symmetry, the amplitude $\langle\phi|\psi\rangle$ is given by

$$\langle\phi|\psi\rangle = \sum_k \langle\phi|\chi_k\rangle\langle\chi_k|\psi\rangle$$

$$= \sum_k \langle\chi_k|\phi\rangle^* \langle\chi_k|\psi\rangle$$

where the states $|\psi\rangle$ and $|\phi\rangle$ are labelled by the measurement outcomes ψ and ϕ. We shall generalize this formula to define the amplitudes between general states defined as normalized vectors in $\mathbb{C}^n$.

We define the amplitude $\langle\phi|\psi\rangle$ between two general states $[\psi]$ and $[\phi]$ (in the χ space) by

$$\langle\phi|\psi\rangle = \sum_k \left(\phi^k\right)^* \psi^k$$

where ψ^k and ϕ^k are the complex components of $[\psi]$ and $[\phi]$ respectively.

It is trivial to see that the amplitude $\langle\phi|\psi\rangle$ is given by

$$\langle\phi|\psi\rangle = [\phi]^\dagger [\psi] = (\phi, \psi)$$

Thus, we can rewrite the definition of probability amplitude as follows:

- *The probability amplitude $\langle\phi|\psi\rangle$ for an event characterized by an initial state $[\psi]$ and final state $[\phi]$ is given by the inner product (ϕ, ψ) between the states*[10].

One should note that the components ψ^i of a state vector $[\psi]^\chi$ have already been defined to be the *amplitudes* $\langle\chi_i|\psi\rangle$. So, in order that the definition of amplitudes provided above be consistent, the components ψ^i must be equal to the inner products (χ_i, ψ). It is trivial to check that this is indeed the case.

---

[10]It may seem that we should also use some indicator like a superscript χ on the inner product (ϕ, ψ) and write it as $(\phi, \psi)^\chi$ to indicate the space that we are working in. Actually (as we shall show in the next section), the inner product between states do not depend on the choice of the space we choose to work in. Of course, this whole description would have broken down if probabilities were dependent on the choice of the space that is used for the description of states.

Description of observables

In this chapter we shall not be able to directly seek a mathematical description of observables. We do not have the mathematical ingredient ready for that as yet. We will do that in the next chapter. However, since we have seen that we can associate a state with every measurement outcome of an observable, we shall try to characterize an observable by identifying the set of states in $\mathbb{C}^n$ that can be associated with the spectrum of an observable.

Owing to the law of sustained states the amplitudes $\langle\psi_j|\psi_i\rangle$, between states that correspond to measurement outcomes of some observable $\hat{\psi}$, obey the relation

$$\langle\psi_j|\psi_i\rangle = \delta_{ji}$$

It is therefore clear that, in order to respect the law of sustained states, the set of state vectors that correspond to the spectrum of an observable must be orthonormal. Moreover, this set must have exactly n elements since we have assumed (see chapter 2, footnote 25) that the spectra of all observables have the same number of elements. Now, we have seen in the problem set that every set of n orthonormal vectors in $\mathbb{C}^n$ comprises an orthonormal basis. Thus the set of state vectors that correspond to the spectrum of an observable *necessarily* constitutes an orthonormal basis. To provide a formal characterization of an observable, we assert that this condition is also *sufficient*. Thus we move to characterize a quantum observable as follows:

- *Every orthonormal basis in $\mathbb{C}^n$ corresponds to some observable such that every member of the spectrum of the observable is associated with a unique member of the orthonormal basis.*

Note that this assumption provides us with only a partial characterization of an observable. We say *partial* because it does not tell us anything about the spectrum of the observable. In the next chapter when we develop the complete characterization of a quantum mechanical observable we shall see how the information about the spectrum is encoded.

In the next subsection, we shall demonstrate how our definitions reproduce the features of QM that we have learnt in the last chapter. But before we get on with that task, let us make sure that we have not lost sight of the whole point of this exercise. We are out to construct a formal description of QM. Recall from the first chapter that in the formal description of ATOPA we wrote down the theory in a language that would make sense even to someone who had never heard of apples but, nonetheless, knew the necessary mathematics to perform analysis with the formal theory. Here, we are trying to describe QM such that it makes sense to anyone who *knows* the necessary vector algebra (of $\mathbb{C}^n$) even if she has no clue about the physics. The construction, thus far, of our formal description has been summarized in the Table 3.1.

TABLE 3.1: Summary of the construction of formal definitions so far. All state vectors have been assumed to be in some definite space, say the χ space.

Motivation	Definition
$\{[\psi]$ is a state$\} \implies \{[\psi]$ is a normalized vector in $\mathbb{C}^n\}$	$\{[\psi]$ is a normalized vector in $\mathbb{C}^n\} \Leftrightarrow \{[\psi]$ is a state$\}$
$\{\langle\phi_j\|\psi_i\rangle$ is an amplitude for $[\psi_i] \to [\phi_j]\} \implies \{\langle\phi_j\|\psi_i\rangle$ is the inner product (ϕ_j, ψ_i) of the normalized vectors $[\phi_j]$ and $[\psi_i]\}$	$\{(\phi, \psi)$ is an inner product of normalized vectors $[\phi]$ and $[\psi]\} \Leftrightarrow \{(\phi, \psi)$ is the amplitude $\langle\phi\|\psi\rangle$ for $[\psi] \to [\phi]\}$
$\{\hat{\psi}$ is an observable with spectrum $\psi\} \implies \{$There exists an orthonormal basis Ψ in $\mathbb{C}^n$ associated with the spectrum $\psi\}$	$\{\Psi$ is an orthonormal basis in $\mathbb{C}^n\} \Leftrightarrow \{$There exists an observable $\hat{\psi}$ such that its spectrum ψ is associated with the orthonormal basis $\Psi\}$

Identification of the laws

In this section we shall demonstrate how the mathematical structure of $\mathbb{C}^n$ provides a description of QM[11]. To this end, we shall look for the algebraic relationships in $\mathbb{C}^n$ that resemble the laws of QM. More concretely, the purpose of this section is to show that

if

1. *we define normalized vectors in $\mathbb{C}^n$ to be quantum states,*

2. *we define inner products between ordered pairs of normalized vectors in $\mathbb{C}^n$ to be the probability amplitudes for the corresponding events, and*

3. *we associate orthonormal bases in $\mathbb{C}^n$ with quantum observables such that every member of any such basis corresponds to a unique member of the spectrum of the associated observable,*

then

the laws of QM as laid down in the previous chapter manifest themselves in the properties of $\mathbb{C}^n$ (endowed with addition of vectors, multiplication of vectors by scalars and inner products).

In order to demonstrate this, we shall first derive a formula for inner products in $\mathbb{C}^n$.

---

[11]It is, of course, not the full formal description of QM but only the first layer in its construction.

If we have two vectors $[\phi]$, $[\psi]$ and an orthonormal basis $\{[\xi_k]\,;k=1,2,\ldots,n\}$ in $\mathbb{C}^n$, then the inner product (ϕ,ψ) between the two vectors is given by

$$(\phi,\psi) \;=\; \sum_{k=1}^{n}(\phi,\xi_k)\,(\xi_k,\psi)$$

To prove this, observe that the vectors $[\phi]$ and $[\psi]$ in $\mathbb{C}^n$ can be written as

$$[\phi] \;=\; \sum_{m}\bar{\phi}^m\,[\xi_m]$$

$$[\psi] \;=\; \sum_{l}\bar{\psi}^l\,[\xi_l]$$

where $\bar{\phi}^m$ and $\bar{\psi}^l$ are appropriate scalars, and all summation indices have been assumed to run from 1 to n. The inner product (ϕ,ψ) will then be given by

$$\begin{aligned}
(\phi,\psi) \;=\; [\phi]^\dagger\,[\psi] \;&=\; \sum_{l}\sum_{m}\left(\bar{\phi}^m\right)^*\bar{\psi}^l\left([\xi_m]^\dagger\,[\xi_l]\right)\\
&=\; \sum_{l}\sum_{m}\left(\bar{\phi}^m\right)^*\bar{\psi}^l\delta_{ml}\\
&=\; \sum_{l}\left(\bar{\phi}^l\right)^*\bar{\psi}^l
\end{aligned}$$

Now let us see what the coefficients $\bar{\psi}^l$ and $\bar{\phi}^l$ are. We have

$$[\psi] \;=\; \sum_{l}\bar{\psi}^l\,[\xi_l]$$

The inner product (ξ_m,ψ) will given by

$$\begin{aligned}
(\xi_m,\psi) \;=\; [\xi_m]^\dagger\,[\psi] \;&=\; \sum_{l}\bar{\psi}^l\left([\xi_m]^\dagger\,[\xi_l]\right)\\
&=\; \sum_{l}\bar{\psi}^l\delta_{ml}\\
&=\; \bar{\psi}^m
\end{aligned}$$

Similarly $(\xi_m,\phi)=\bar{\phi}^m$. Thus, we finally have

$$\begin{aligned}
(\phi,\psi) \;&=\; \sum_{l}\left(\bar{\phi}^l\right)^*\bar{\psi}^l\\
&=\; \sum_{l}(\xi_l,\phi)^*\,(\xi_l,\psi)\\
&=\; \sum_{l}(\phi,\xi_l)\,(\xi_l,\psi)
\end{aligned}$$

q.e.d

Now let us take up the laws of QM one by one to see how they are embedded in the algebraic structure of $\mathbb{C}^n$.

Law of definite probabilities This follows from the fact that to every ordered pair of vectors there exists a well defined complex number given by the inner product. The square of the modulus of this complex number is obviously a positive semidefinite real number. For a pair of normalized vectors, this number lies between zero and one[12]. Further, it is easy to see that for a normalized vector, the inner product formula derived above reduces to the totality condition when we set a self inner product to unity.

Exchange symmetry Exchange symmetry follows trivially from the conjugate symmetry property of inner products.

Superposition principle To see how this comes about, recall that the superposition principle reads

$$\langle \phi_j | \psi_i \rangle = \sum_k \langle \phi_j | \xi_k \rangle \langle \xi_k | \psi_i \rangle$$

where the states $|\xi_k\rangle$ are labelled by the elements of the spectrum of some arbitrary observable $\hat{\xi}$ and k runs over its entire spectrum. It is immediately evident that the inner product formula that we have derived above will reduce to the superposition principle when $[\phi]$ and $[\psi]$ are normalized vectors so that the scalar products appearing in the formula can be interpreted as amplitudes:

$$\langle \phi | \psi \rangle = \sum_l \langle \phi | \xi_l \rangle \langle \xi_l | \psi \rangle$$

noting that the vectors $[\xi_l]$ have already been assumed to be normalized[13].

A change in the notation of inner products

Before we quit this section let us note that the inner products between arbitrary vectors, irrespective of whether they constitute quantum states or

---

[12]This follows from Cauchy-Schwarz inequality:

$$|(\phi, \psi)| \leq |\phi| \, |\psi|$$

for all $[\phi], [\psi] \in \mathbb{C}^n$. Actually, it applies to more general spaces that are generalizations of $\mathbb{C}^n$. We will see such spaces shortly.

[13]The law of sustained states follows from the fact that the vectors that are associated with the members of the spectrum of an observable are, by assumption, members of an orthonormal set. But we do not really need to include this in our demonstration, since we have pointed out in the last chapter that this is no longer an independent law of QM.

not (i.e., irrespective of whether they are normalized or not), bear a strong resemblance with probability amplitudes. To wit

1. exchange symmetry: this is just the conjugate symmetry property in the definition of inner product

$$(\phi, \psi) \;=\; (\psi, \phi)^*$$

2. superposition principle: this is manifest in the formula for inner products that we have proved in the beginning of the last subsection

$$(\phi, \psi) \;=\; \sum_l (\phi, \xi_l) (\xi_l, \psi)$$

for every orthonormal basis $\{[\xi_l] \,;\, l = 1, 2, \ldots, n\}$.

Thus we see that, although not all inner products are amplitudes, they virtually mimic the properties of amplitudes. So, from here on we shall use the same notation for amplitudes and inner products and denote all inner products[14] (ϕ, ψ) by $\langle \phi | \psi \rangle$.

Problems

1. Which of the following vectors can qualify as a state in QM?
   a) $(1, 1, 1)^T$ b) $(1/\sqrt{3}) (1 + i, 1)^T$ c) $(1, 0, 0, 0)^T$ d) $(1 + i, 1 - i)^T$
   Multiply the vectors that do not qualify as states by appropriate scalars so that they do qualify as states.

2. Two states $|\psi\rangle$ and $|\phi\rangle$ are given in the χ space as

$$[\psi]^\chi = \frac{1}{5\sqrt{2}} \begin{bmatrix} 3 + 4i \\ 4 + 3i \end{bmatrix} \quad \text{and} \quad [\phi]^\chi = \frac{1}{5\sqrt{2}} \begin{bmatrix} 4 - 3i \\ 3 - 4i \end{bmatrix}$$

   Find out the amplitudes and probabilities for the following events:
   a) $|\psi\rangle \to |\phi\rangle$ b) $|\phi\rangle \to |\psi\rangle$ c) $|\psi\rangle \to |\chi_1\rangle$ d) $|\chi_2\rangle \to |\phi\rangle$

3. Check whether a measurement on a state $(1/\sqrt{5}) (1, 2)^T$ can lead to the state $(1/\sqrt{5}) (-2, 1)^T$.

4. Consider the set of vectors $\left\{ (1, 0, 0)^T, (0, 1, 0)^T, \frac{1}{\sqrt{2}} (0, 1, 1)^T \right\}$ in $\mathbb{C}^3$. Is it possible that the given set of vectors correspond to the complete set of states associated with the spectrum (comprising three elements) of some observable?

---

[14] After reading this chapter, please see the section on *braket* notation in appendix 'B' for a deeper motivation of this notation.

5. Consider the two vectors

$$[\psi]^\chi = \frac{1}{\sqrt{3}} \begin{bmatrix} 1+i \\ 1 \end{bmatrix} \quad \text{and} \quad [\phi]^\chi = \frac{1}{\sqrt{4}} \begin{bmatrix} 1+i \\ 1-i \end{bmatrix}$$

(a) Show that the vectors can represent quantum states.

(b) Investigate if they can represent states corresponding to the same observable with different outcomes.

(c) What would be the probability of the event $[\psi] \to [\phi]$?

6. Prove Cauchy-Schwarz inequality: For any two vectors $[A]$ and $[B]$ in $\mathbb{C}^n$

$$|(A,B)| \leq |A|\,|B|$$

7. Our definition of a quantum state means that to every normalized vector there will always exist an observable such that the vector can be associated with one of its outcomes. Mathematically, it is not difficult to see why this is so. Given a normalized vector one merely has to construct an orthonormal basis that includes the vector[15]. Justify that this is always possible.

Vector Spaces and Inner Product Spaces

In the previous section we have assumed that states in the χ space live in $\mathbb{C}^n$. In fact, all vectors of $\mathbb{C}^n$ will carry the signature of the χ space since any vector can be normalized to describe a state vector. Obviously, there is nothing special about the χ space. We can choose to use any other space that we fancy. If we chose some other space, say ξ, we would along the same lines, again use $\mathbb{C}^n$ to host the state vectors (only this time, in the ξ space). In this section we shall try to see the mathematical relationship between these different descriptions. It is clearly essential that all such descriptions should reproduce the same physics (i.e., the same *facts*), and therefore they should somehow integrate into some common unified framework. Such a framework will indeed be seen to exist. However, to see this we will have to understand the algebraic structure of $\mathbb{C}^n$ from a more general and abstract point of view.

[15] However, it is neither easy nor essential to physically make sense of observables associated with arbitrary orthonormal bases.

Different representations of a vector in $\mathbb{C}^n$

Consider a basis $\xi = \{[\xi_k]; k = 1, 2, \ldots, n\}$ in $\mathbb{C}^n$. Then an arbitrary vector $[\psi]$ in $\mathbb{C}^n$ can be written as a linear combination

$$[\psi] = \sum_{k=1}^{n} \bar{\psi}^k [\xi_k]$$

We can associate a *unique* n-tuple of coefficients $\bar{\psi}^k$ with every vector $[\psi]$ in $\mathbb{C}^n$. We call this n-tuple the **representation** of the vector $[\psi]$ in the ξ basis (or simply the ξ representation of $[\psi]$). The coefficients $\bar{\psi}^k$ are called the **coordinates** of the representation. We can write the ξ representation of the vector $[\psi]$ as a column matrix as well

$$[\psi]^\xi = \begin{bmatrix} \bar{\psi}^1 \\ \bar{\psi}^2 \\ .. \\ .. \\ \bar{\psi}^n \end{bmatrix}$$

Clearly, the representation $[\psi]^\xi$ is also an element of $\mathbb{C}^n$. In fact, the collection of ξ representations of all vectors in $\mathbb{C}^n$ is identical to the set $\mathbb{C}^n$ (i.e., it gives us another *copy* of $\mathbb{C}^n$). This means, that for every basis ξ, there is a one-to-one mapping from $\mathbb{C}^n$ onto itself.

We shall work exclusively with orthonormal bases. As we have seen before, if a basis ξ is orthonormal, the expansion coefficients are given by the inner product so that

$$\bar{\psi}^k = \langle \xi_k | \psi \rangle$$

Incidentally, in the standard basis $\chi = \{[\chi_k]; k = 1, 2, \ldots, n\}$ where $\chi_k^i = \delta_{ik}$, a vector $[\psi]$ is expanded as

$$[\psi] = \sum_{i=1}^{n} \psi^i [\chi_i]$$

That is, the *coordinates* of $[\psi]$ in the standard representation are simply the *components* of $[\psi]$. Thus the χ representation $[\psi]^\chi$ of $[\psi]$ is identical to $[\psi]$:

$$[\psi]^\chi = [\psi]$$

Since the standard basis is also orthonormal, we have

$$\psi^i = \langle \chi_i | \psi \rangle$$

Relationship between representations

It is easy to see that if ζ and ξ are two orthonormal bases, then the coordinates $\langle \zeta_k | \psi \rangle$ and $\langle \xi_k | \psi \rangle$ of a vector $[\psi]$ in the two representations $[\psi]^\zeta$

and $[\psi]^\xi$ are related by[16]

$$\langle\zeta_i|\psi\rangle = \sum_j \langle\zeta_i|\xi_j\rangle\langle\xi_j|\psi\rangle$$

This can be written as a matrix equation

$$[\psi]^\zeta = T_{(\zeta,\xi)}[\psi]^\xi$$

where $T_{(\zeta,\xi)}$ is a $n \times n$ square matrix whose components $T^{ij}_{(\zeta,\xi)}$ are defined by

$$T^{ij}_{(\zeta,\xi)} = \langle\zeta_i|\xi_j\rangle$$

Note that the matrix $T_{(\zeta,\xi)}$ is made up of the column vectors $[\xi_k]^\zeta$

$$T_{(\zeta,\xi)} = \begin{bmatrix} \langle\zeta_1|\xi_1\rangle & \langle\zeta_1|\xi_2\rangle & \cdots & \cdots & \langle\zeta_1|\xi_n\rangle \\ \langle\zeta_2|\xi_1\rangle & \langle\zeta_2|\xi_2\rangle & \cdots & \cdots & \langle\zeta_2|\xi_n\rangle \\ \cdots & & \cdots & \cdots & \cdots \\ \cdots & & \cdots & \cdots & \cdots \\ \langle\zeta_n|\xi_1\rangle & \langle\zeta_n|\xi_2\rangle & \cdots & \cdots & \langle\zeta_n|\xi_n\rangle \end{bmatrix}$$

Owing to the orthonormality of the vectors $[\xi_k]^\zeta$, the matrix $T_{(\xi,\chi)}$ will turn out to be unitary[17]. The matrix $T_{(\zeta,\xi)}$ transforms the ξ representation $[\psi]^\xi$ of a vector to its ζ representation $[\psi]^\zeta$. It is therefore called a ***transformation matrix***. The transformation matrix defines a function from $\mathbb{C}^n$ onto itself. We will denote this function (i.e., the *transformation*) also by $T_{(\zeta,\xi)}$.

Let us now consider the set of ξ representations of all vectors in $\mathbb{C}^n$. Let us call this set $\mathbb{C}^n_\xi$. We have already remarked that this set is identical to $\mathbb{C}^n$. It is only the interpretation of their elements that are different. We can define addition, multiplication by a scalar and inner product on $\mathbb{C}^n_\xi$ in the same way as on $\mathbb{C}^n$. We can similarly define $\mathbb{C}^n_\zeta$. Now we note that the general transformation matrix $T_{(\zeta,\xi)}$ which defines a mapping from $\mathbb{C}^n_\xi$ to $\mathbb{C}^n_\zeta$ *preserves the algebraic structure*. By this we mean that

- *under the transformation $T_{(\zeta,\xi)}$, the image of an arbitrary linear combination of vectors in $\mathbb{C}^n_\xi$ is equal to the same linear combination of the*

---

[16]This relation was actually derived in the last section. Up until that point we were using a slightly different notation for inner products. In that notation this relation would read

$$(\zeta_i,\psi) = \sum_j (\zeta_i,\xi_j)(\xi_j,\psi)$$

[17]It turns out that when the bases are not orthonormal, the transformation matrix $T_{(\xi,\chi)}$ is still made up of the column vectors $[\chi_k]^\xi$, however, $T_{(\xi,\chi)}$ is then no longer unitary. After you have read this chapter, you may check out appendix 'B' for transformation between nonorthonormal bases.

images of the of vectors in $\mathbb{C}^n_\varsigma$

$$\sum_i c_i \, [\psi_i]^\xi \xrightarrow{\;T_{(\varsigma,\xi)}\;} \sum_i c_i \, [\psi_i]^\varsigma$$

where the $c_i's$ are arbitrary complex numbers.

This follows trivially from the relationship between the representations[18]

$$[\psi]^\varsigma = T_{(\varsigma,\xi)} \, [\psi]^\xi$$

In general, when a mapping preserves the algebraic structure, it is called a *homomorphism*. When the homomorphism is one-to-one, it called an *isomorphism*. If we have an isomorphism from a set onto itself, it is called an *automorphism*. Thus, every transformation $T_{(\varsigma,\xi)}$ defines an automorphism on $\mathbb{C}^n$.

Now, the transformation $T_{(\xi,\chi)}$ effects a transformation from the standard representation χ to the representation ξ. Since $[\psi]^\chi = [\psi]$, and the transformation $T_{(\xi,\chi)}$ preserves the algebraic structure, it follows that the representations $[\psi]^\xi$ in $\mathbb{C}^n_\xi$ essentially *mimic* the algebraic behaviour of the vectors $[\psi]$ in the $\mathbb{C}^n$ that we started out with. This, of course, is true for any representation ξ.

We observe further that the unitarity of the transformation matrix $T_{(\xi,\chi)}$ implies that for an arbitrary pair of vectors $[\psi]$ and $[\phi]$

$$([\phi])^\dagger \, [\psi] = \left(T_{(\xi,\chi)} \, [\phi]\right)^\dagger \left(T_{(\xi,\chi)} \, [\psi]\right) = \left([\phi]^\xi\right)^\dagger [\psi]^\xi$$

This means that

the inner product of an arbitrary pair of vectors is invariant under the transformation.

That is, irrespective of whether we use the coordinates in the standard representation (i.e., the components of the vectors) or the coordinates in some other orthonormal representation, the inner product remains unaltered.

Problems

1. Two orthonormal bases in $\mathbb{C}^2$ are given by

$$S = \left\{ \frac{1}{\sqrt{2}} \begin{bmatrix} 1 \\ 1 \end{bmatrix}, \; \frac{1}{\sqrt{2}} \begin{bmatrix} 1 \\ -1 \end{bmatrix} \right\} \qquad \text{and} \qquad S' = \left\{ \frac{1}{5} \begin{bmatrix} 4 \\ -3 \end{bmatrix}, \; \frac{1}{5} \begin{bmatrix} 3 \\ 4 \end{bmatrix} \right\}$$

---

[18]Multiplication of matrices distributes over addition and associates with scalar multiplication of matrices.

(a) Find the transformation matrices $T_{(S,E)}$ and $T_{(S',S)}$ that effects the transformations from the standard basis E to S and from S to S'. Express $T_{(S',E)}$ in terms of $T_{(S,E)}$ and $T_{(S',S)}$.

(b) Demonstrate that the transformation matrices are unitary.

(c) Find out the inner product $\langle \psi | \phi \rangle$ between the vectors $[\psi]^S = (1/\sqrt{5})\,(1,2)^T$ and $[\phi]^{S'} = (1/\sqrt{10})\,(1,3)^T$.

2. The following matrix $T_{(\xi,\eta)}$ transforms coordinates from an orthonormal basis η to a basis ξ in $\mathbb{C}^3$:

$$T_{(\xi,\eta)} = \frac{1}{\sqrt{2}} \begin{bmatrix} 1 & 1 & 0 \\ -i & i & 0 \\ 0 & 0 & \sqrt{2}i \end{bmatrix}$$

(a) Write down the elements of the basis ξ in terms of the elements of η.

(b) Can you check, simply by looking at T, whether the basis ξ is orthonormal?

(c) If the vector $(1,1,1)^T$ is a representation in the ξ basis, express it in the η basis.

3. Show that the transformation matrix taking representations from one orthonormal basis to another is necessarily unitary. Prove that the converse is also true: every unitary matrix can be looked upon as a transformation matrix that effects a transformation between two orthonormal bases.

Abstract spaces

Recall that ordinary displacement like vectors in our usual 3-dimensional space are represented as ordered triplets of real numbers (i.e., as elements of $\mathbb{R}^3$). These representations (which can be written as real, 3-dimensional column matrices) *mimic* the behaviour of the displacement vectors they represent. They have the same algebraic structure under addition of column matrices and multiplication of column matrices by scalars (real numbers) as displacement like vectors under vector addition and multiplication of a vector by a scalar (defined by parallelogram law and length scaling, respectively). Obviously, this algebraic structure is preserved if we go from one representation (coordinate system) to another (say, a rotated coordinate system). Now, in the previous subsection we have seen that the algebraic structure of $\mathbb{C}^n$ is also preserved as we go from one of its representations to another. We would like to explore the possibility whether we can imagine an underlying set of *objects* (analogous to displacement like vectors) to which $\mathbb{C}^n$ is a set of representations[19]. We could call these underlying objects, vectors, as well. This will enable us to speak of vectors in a representation independent way. We could, in some sense, regard them as more *fundamental* than their representations.

Let us then imagine an *abstract space* that has an algebraic structure exactly like $\mathbb{C}^n$. We are essentially contemplating a set of symbols that can be put into one-to-one correspondence with the vectors of $\mathbb{C}^n$. We define definite rules of combining the symbols and scalars (i.e., complex numbers) to yield other symbols in the set (to correspond to addition and scalar multiplication of vectors) or combining pairs of symbols to yield a scalar (to correspond to inner product of vectors) in such a way that the algebra of such combinations is exactly the same as the algebra of $\mathbb{C}^n$. Let me define this more concretely.

Let us consider a set V and call its elements, **vectors**. We will denote vectors as $|\psi\rangle, |\phi\rangle, \ldots$, etc. Let us define an operation that associates a vector to every ordered pair $(|\psi\rangle, |\phi\rangle)$ of vectors in V. We choose to denote this vector by $|\psi\rangle + |\phi\rangle$ and call this operation *"addition of vectors"*. Let us define another operation that associates a vector, denoted by $c\,|\psi\rangle$ to every pair $\{c, |\psi\rangle\}$ consisting of a complex number c (to be called a **scalar** again) and a vector $|\psi\rangle$ in V. We will call this operation *"multiplication of a vector by a scalar"*. Now let us require the operations, addition of two vectors and multiplication of a vector by a scalar, to obey the properties that were listed for $\mathbb{C}^n$ in the first section of this chapter:

1. Associativity of addition of vectors: For all vectors $|\alpha\rangle$, $|\beta\rangle$, $|\gamma\rangle$

$$(|\alpha\rangle + |\beta\rangle) + |\gamma\rangle = |\alpha\rangle + (|\beta\rangle + |\gamma\rangle)$$

2. Existence of Additive Identity: There exists a vector $|\,\rangle$, that is called the **additive identity**, such that for any vector $|\alpha\rangle$

$$|\alpha\rangle + |\,\rangle = |\,\rangle + |\alpha\rangle = |\alpha\rangle$$

3. Existence of Additive Inverse: For every vector $|\alpha\rangle$ there exists a vector $-\,|\alpha\rangle$, that is called the **additive inverse** of $|\alpha\rangle$, such that[20]

$$|\alpha\rangle + (-\,|\alpha\rangle) = (-\,|\alpha\rangle) + |\alpha\rangle = |\,\rangle$$

4. Commutativity of addition of vectors: For all vectors $|\alpha\rangle$, $|\beta\rangle$

$$|\alpha\rangle + |\beta\rangle = |\beta\rangle + |\alpha\rangle$$

---

[19]This exercise is an important example in QM of the process of *abstraction and generalization* that we discussed in the first chapter.

[20]Incidentally, an algebraic system that obeys these first three properties is called a **group**, and when a group obeys the next property (i.e., commutativity), it is called a **commutative group** (also called an **Abelian group**). *Groups* are profoundly important algebraic structures in all of physics and mathematics.

5. Compatibility of multiplication of scalars with multiplication of a vector by a scalar: For every vector $|\alpha\rangle$ and all scalars a, b

$$a\left(b|\alpha\rangle\right) = (ab)\,|\alpha\rangle$$

6. Distributivity of scalar multiplication of a vector over addition of vectors: For all vectors $|\alpha\rangle$ and $|\beta\rangle$ and every scalar a

$$a\left(|\alpha\rangle + |\beta\rangle\right) = a|\alpha\rangle + a|\beta\rangle$$

7. Distributivity of scalar multiplication of a vector over addition of scalars: For every vector $|\alpha\rangle$ and all scalars a, b

$$(a+b)\,|\alpha\rangle = a|\alpha\rangle + b|\alpha\rangle$$

8. Equality of identity of scalar multiplication and multiplicative identity of scalars: For every vector $|\alpha\rangle$

$$1|\alpha\rangle = |\alpha\rangle$$

The set V endowed with these properties is said to form a ***vector space over the field of complex numbers***[21].

Now let us define another function that assigns a scalar denoted by $\langle\phi|\psi\rangle$ to every ordered pair $(|\phi\rangle, |\psi\rangle)$ of vectors in V. Let us call this function an *inner product of vectors*. We require this function to obey the properties that were listed for inner product of vectors in $\mathbb{C}^n$:

1. Conjugate Symmetry: For all vectors $|\alpha\rangle$, $|\beta\rangle$

$$\langle\alpha|\beta\rangle = \langle\beta|\alpha\rangle^{*}$$

2. Linearity: For all vectors $|\alpha\rangle$, $|\beta\rangle$, $|\gamma\rangle$ and all scalars b, c

$$\langle\alpha|,(b\,|\beta\rangle + c\,|\gamma\rangle) = b\,\langle\alpha|\beta\rangle + c\,\langle\alpha|\gamma\rangle$$

3. Positive-definiteness: For any vector $|\alpha\rangle$

$$\langle\alpha|\alpha\rangle \geq 0 \quad \text{and} \quad \langle\alpha|\alpha\rangle = 0 \quad \text{if and only if} \quad |\alpha\rangle = |\,\rangle$$

---

[21]Linear algebra, which provides the main mathematical basis for QM, is essentially the study of vector spaces. A firm grasp of this very important branch of mathematics is actually indispensable for almost all areas of physics and mathematics. Although we provide the essentials required for this discourse, we recommend the comprehensive treatise by P.R.Halmos (Halmos 1987) for acquiring a thorough background on this subject.

The vector space V is now said to form an ***inner product space over the complex field***[22].

Note that in the above definitions the names, *addition* and *multiplication* that we have used for the functions are *not* the usual addition and multiplication of numbers or matrices. We have merely borrowed the names[23].

It is easy to imagine that one can define a vector space and inner product space with real numbers as scalars. Such a space is often called an *Euclidean space*[24]. An n-dimensional Euclidean space is denoted by $\mathbb{R}^n$. Incidentally, an inner product space over a complex field (that we have just defined) is called an *unitary* space. There are myriads of examples of vector spaces and inner product spaces other than $\mathbb{R}^n$ and $\mathbb{C}^n$. However, it is easy to see that *every n-dimensional real vector space is isomorphic to $\mathbb{R}^n$ and every n-dimensional complex vector space is isomorphic to $\mathbb{C}^n$.*

All definitions that were laid down for $\mathbb{C}^n$ in the first section of this chapter carry over naturally to abstract spaces. The ***norm*** of a vector is the positive square root of the inner product of the vector with itself. Vectors which have unit norm are said to be ***normalized vectors***. If two vectors have a vanishing inner product, they are called ***orthogonal*** vectors. A set comprising mutually orthogonal vectors is called an ***orthogonal set***. If an orthogonal set consists of normalized vectors, it is called an ***orthonormal set***. A ***linear combination*** of a set of vectors is a sum of scalar multiples of the vectors. When an arbitrary vector in a vector space can be expressed as a linear combination of the elements of some subset of the vector space, we say that the subset forms a ***complete set***. The set of all possible linear combinations of a subset is called the ***linear span*** of the subset. If it is impossible to express any of the vectors of a set as a linear combination of other vectors in the set, we say that the set is a ***linearly independent set***. A set which is complete and linearly independent is said to form a ***basis***. A basis which is orthonormal is naturally called an ***orthonormal basis***. It turns out that

---

[22]Essentially, we have defined three mappings or functions: $V \times V \longrightarrow V$, $\mathbb{C} \times V \longrightarrow V$ and $V \times V \longrightarrow \mathbb{C}$. The values of the functions are denoted respectively by $|\psi\rangle + |\phi\rangle$, $c|\psi\rangle$ and $\langle \phi|\psi\rangle$:

$$(|\psi\rangle, |\phi\rangle) \mapsto |\psi\rangle + |\phi\rangle \qquad (c, |\psi\rangle) \mapsto c|\psi\rangle \qquad (|\psi\rangle, |\phi\rangle) \mapsto \langle \phi|\psi\rangle$$

and they have been defined to obey the properties that we have listed. Here '$\times$' is the Cartesian product. Incidentally, a function like $V \times V \longrightarrow V$ is also called a binary operation on V. For those unfamiliar, the symbol '$\mapsto$' means "maps to".

[23]In the first few problems of the next problem set, you will see how the structure of vector spaces and inner product spaces can show up in completely different contexts and forms.

[24]If one uses real instead of complex scalars one would have, what are called, vector spaces and inner product spaces defined over the real field. Actually, such algebraic structures can be defined on arbitrary fields where *field* is an abstraction of the familiar algebraic systems of real and complex numbers (i.e., the set of real or complex numbers with addition and multiplication defined on them).

every basis of a vector space has exactly the same number of elements, and this number is called the ***dimension*** of the vector space[25].

Incidentally, if S be a subset of a vector space V, and if S forms a vector space over the same field under the same addition and scalar multiplication operations under which V forms a vector space, then S is called a *subvector space*, or simply, a ***subspace***[26].

Let us now see how we can retrieve $\mathbb{C}^n$ from our abstract space. We choose some orthonormal basis $\xi = \{|\xi_k\rangle; k = 1, 2, \ldots, n\}$ in V. Then an arbitrary vector $|\psi\rangle$ in V can be written as

$$|\psi\rangle = \sum_{k=1}^{n} |\xi_k\rangle \langle \xi_k | \psi \rangle$$

The ξ ***representation*** of $|\psi\rangle$ is defined as the n-tuple of expansion coefficients

$$[\psi]^\xi = \begin{bmatrix} \langle \xi_1 | \psi \rangle \\ \langle \xi_2 | \psi \rangle \\ \\ \langle \xi_n | \psi \rangle \end{bmatrix}$$

The set of ξ representations of all vectors $|\psi\rangle$ in V will comprise the ξ representation space[27] $\mathbb{C}_\xi^n$. It is trivial to see that $\mathbb{C}_\xi^n$ is identical to $\mathbb{C}^n$. Thus, vectors that were defined in the *different spaces* are now treated as representations of the same vector in *different bases*. Incidentally, the set of representations of the basis vectors $|\xi_k\rangle$ in their own basis ξ will constitute the standard basis in the ξ representation space $\mathbb{C}_\xi^n$:

$$|\xi_k\rangle = \sum_{j=1}^{n} \delta_{jk} |\xi_j\rangle$$

$$\implies \quad \xi_k^j = \delta_{jk}$$

where ξ_k^j is the j-th component of $[\xi_k]^\xi$.

Problems

1. Show that the ordinary 3-vectors (directed line segments in 3-dimensional space) form an inner product space with real scalars under

---

[25] See problem '9' in the problem set that follows.

[26] Please see appendix 'B' for a few easy to prove but important theorems on subspaces. We will need them later in the book.

[27] Such a space will often be referred to as a ξ space in short.

addition of vectors (parallelogram or triangle law), multiplication of a vector by a scalar (scaling) and dot product of vectors.

2. Consider the set of 2×2, traceless, Hermitian matrices.

   (a) Show that they form a vector space under addition of matrices and multiplication of matrices by a (real) scalar.

   (b) What will be the dimension of this vector space?

   (c) If we define a function that associates every pair of such matrices (A, B) with a real number $\Phi(A, B)$ according to the rule:

$$\Phi(A, B) = \frac{1}{4} \text{Tr}(AB + BA)$$

   establish that the function qualifies as an inner product.

   (Hint: Write the general form of the matrices in terms of real scalars. For example, write the traceless Hermitian matrices in terms of three real numbers x, y, and z as

$$\begin{bmatrix} z & x - iy \\ x + iy & -z \end{bmatrix}$$

   Now observe that the problem can be identified with the previous problem if you identify every such matrix with the ordered triplet (x, y, z). You may try to identify (read off) a basis, and proceed from there.)

3. Show that the set of n-th degree (univariate) polynomials (with real coefficients) $\mathcal{P}_n(t)$ form a vector space under addition (of polynomials) and multiplication (of polynomials) by a scalar.

4. Let S be a set of positive real numbers. We define two operations $\oplus$ and $\odot$ on S by

$$\begin{aligned} x \oplus y &= xy \\ x \odot r &= x^r \end{aligned}$$

   for all x, y in S and all real numbers r. Show that the set S forms a vector space over the real field under the operations[28] $\oplus$ and $\odot$.

5. Consider, once again, the ordinary 3-vectors of the first problem.

---

[28] Note how, ordinary multiplication plays the role of vector addition while multiplication by a scalar is given by exponentiation. In abstract algebra, the terms *addition* and *multiplication* are just names. Notice that the objects that constitute the vector spaces in this and the preceding three problems are actually completely different *animals*. Nevertheless, they have many common features by virtue of only a few fundamental properties (namely, the defining conditions of a vector space) they share.

(a) Show that any set of three non-coplanar vectors will form a basis in the above vector space.

(b) The set $S = \{\hat{x}, \hat{y}, \hat{z}\}$ forms an orthonormal basis where $\hat{x}, \hat{y}$ and $\hat{z}$ are the unit vectors along the axes of a Cartesian coordinate system. If the $x - y$ plane of the Cartesian system S is rotated by some angle θ about the z axis to yield a new orthogonal coordinate system S', construct the transformation matrix R that connects S and S'. Show that the transformation matrix R is an orthogonal matrix.

(c) Argue that any orthogonal matrix will always be a transformation matrix that effects a transformation from one orthonormal basis to another (i.e., from one orthogonal co-ordinate system to another[29]).

6. Let $\chi = \{|\chi_1\rangle, |\chi_2\rangle\}$ be an orthonormal basis for some inner product space V and $\xi = \{|\xi_1\rangle, |\xi_2\rangle\}$ be another set contained in V, where

$$|\xi_1\rangle = \frac{-i}{\sqrt{2}}|\chi_1\rangle + \frac{1}{\sqrt{2}}|\chi_2\rangle \qquad \text{and} \qquad |\xi_2\rangle = \frac{1}{\sqrt{2}}|\chi_1\rangle + \frac{-i}{\sqrt{2}}|\chi_2\rangle$$

(a) Show that ξ forms an orthonormal basis in V as well.

(b) Construct the transformation matrix $T_{(\xi,\chi)}$.

(c) Determine the coordinates of the vector $|\psi\rangle = (1/5)(3|\chi_1\rangle + 4|\chi_2\rangle)$ in the ξ basis using $T_{(\xi,\chi)}$.

(d) If a vector is given by $|\phi\rangle = (1/5)(4|\xi_1\rangle - 3|\xi_2\rangle)$, compute the inner product $\langle\phi|\psi\rangle$.

7. If $B = \{|\beta_1\rangle, |\beta_2\rangle, |\beta_3\rangle, |\beta_4\rangle\}$ forms a basis in V and $|\alpha\rangle = c_2|\beta_2\rangle + c_4|\beta_4\rangle$ with $c_2, c_4 \neq 0$, show that the set $B' = \{|\beta_1\rangle, |\alpha\rangle, |\beta_3\rangle, |\beta_4\rangle\}$ also forms a basis.

(Hint: In the expansion of any vector in terms of the vectors $|\beta_i\rangle$, one can always replace $|\beta_2\rangle$ or $|\beta_4\rangle$ in terms of $|\alpha\rangle$ to establish completeness. The linear independence of the new set is a consequence of the linear independence of the old set.

It is straightforward to generalize the analysis to any finite dimensional vector space. In the general setting this gives what is known as the *replacement theorem*: *If $B = \{|\alpha_i\rangle; i = 1, 2, \ldots, n\}$ be a basis in a vector space V and $|\psi\rangle$ be a vector given by*

$$|\psi\rangle = \sum_{i=1}^{n} c_i |\alpha_i\rangle$$

where c_is are scalars, then

$$B' = \{|\alpha_1\rangle, |\alpha_2\rangle, \ldots, |\alpha_{k-1}\rangle, |\psi\rangle, |\alpha_{k+1}\rangle, \ldots, |\alpha_n\rangle\}$$

---

[29]Check out the more general theorem in an abstract setting in appendix 'B'.

forms a basis as well, provided that $c_k \neq 0$.)

8. If $B = \{|\alpha_1\rangle, |\alpha_2\rangle, |\alpha_3\rangle\}$ forms a basis in a vector space V, show that the set $B' = \{|\beta_1\rangle, |\beta_2\rangle, |\beta_3\rangle\}$ also forms a basis of the vector space in V where

$$|\beta_1\rangle = |\alpha_1\rangle + |\alpha_2\rangle \quad |\beta_2\rangle = |\alpha_1\rangle + 2|\alpha_2\rangle + 3|\alpha_3\rangle \quad \beta_3 = 2|\alpha_1\rangle + 3|\alpha_2\rangle + 4|\alpha_3\rangle$$

(Hint: Use the argument of problem (7) sequentially, to replace the members of B by those of B'. Inspect that B' is linearly independent, and note that the process of replacement will not *fail* owing to this.
Again, the procedure is easily extended to arbitrary finite dimensional vector spaces. The general result is: *if a basis of a vector space V has n elements then every set of n linearly independent vectors in V will form a basis*[30].)

9. Establish that if one basis in a vector space V has n elements, then every basis of V must also have exactly n elements[31]. Hence, argue that if B is a basis in a vector space V, then it is the smallest complete set and the largest linearly independent set in V.

10. Show that the dependence of the inner product on the first argument is antilinear: using the notation $\langle\alpha|\beta\rangle \equiv (|\alpha\rangle, |\beta\rangle)$, demonstrate that

$$(c_1|\alpha\rangle + c_2|\beta\rangle, |\gamma\rangle) = c_1^*(|\alpha\rangle, |\gamma\rangle) + c_2^*(|\beta\rangle, |\gamma\rangle)$$

where $|\alpha\rangle, |\beta\rangle, |\gamma\rangle$ are arbitrary vectors and c_1, c_2 are arbitrary scalars.

11. Prove that, given a *set* of linearly independent vectors, one can construct an orthonormal set (of the same cardinality) by taking appropriate linear combinations of the given vectors.

(Hint: Imagine three non-coplanar vectors which are not mutually orthogonal. Now construct three orthonormal vectors by taking appropriate linear combinations. Generalize the algebra of the proof to your abstract inner product space. If you succeed, you will have proved the Gram-Schmidt orthogonalization theorem[32].)

12. Using the Gram-Schmidt orthogonalization process, construct an orthogonal basis from the basis vectors: $(0, 2, -1), (2, 0, 1), (2, 2, 0)$ in $\mathbb{R}^3$.

13. Prove that every n-dimensional real vector, space is isomorphic to $\mathbb{R}^n$, and every n-dimensional complex vector space is isomorphic to $\mathbb{C}^n$.

---

[30] The general proof has been provided in the appendix. However, we would strongly encourage you to do it yourself.

[31] Recall that you have already proved this in the first problem set for $\mathbb{C}^n$. Here you are required to frame the argument in a more abstract setting.

[32] If you do not succeed, please look up in appendix 'B'.

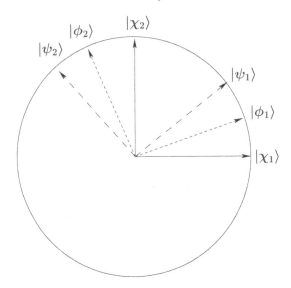

FIGURE 3.1: The state space for a 2-dimensional quantum system visualized as a 2-dimensional real vector space. States lie on the unit circle. A measurement of $\hat{\phi}$ on $|\psi_1\rangle$ leading to an outcome ϕ_2 produces a final state $|\phi_2\rangle$. This visualization is, however, not precisely correct since unit normed state vectors are not unique and defined only up to global phases in a complex vector space.

Postulates of Quantum Mechanics - Version 1

It is now easy to guess the appropriate mathematical framework for the description of QM that will treat states in different spaces in a unified way. We shall have to formulate QM using an abstract inner product space instead of $\mathbb{C}^n$ for the description of the *state space* (see Figure 3.1). We will lay down the formulation as a set of postulates. In the course of this book, these postulates will evolve as we introduce new features of QM and new ingredients of description. It is our purpose to demonstrate how this evolution comes about. Let us then write down the first version of the postulates.

1. Every system is associated with an inner product space. States of the system are the normalized vectors belonging to this space.

2. Every observable of the system is associated with an orthonormal basis of the state space. The members of the measurement spectrum of an observable are in one to one correspondence with the members of the orthonormal basis associated with the observable[33].

---

[33]Note that we are still not in a position to say what an observable *is* in terms of a concrete, mathematical ingredient of our formal description. In particular, these postulates

3. If a measurement of an observable $\hat{\psi}$ yields an outcome ψ_i, soon after the measurement, the new state becomes $|\psi_i\rangle$ - the state associated with the outcome[34] ψ_i.

4. The probability $P[\psi_i \to \phi_j]$ of the event $\psi_i \to \phi_j$ of getting an outcome ϕ_j upon measurement of an observable $\hat{\phi}$ on a state $|\psi_i\rangle$ is given by the *Born rule*:

$$P[\psi_i \to \phi_j] \;=\; |\langle\phi_j|\psi_i\rangle|^2$$

where the scalar product $\langle\phi_j|\psi_i\rangle$ is called the probability amplitude of the event $\psi_i \to \phi_j$.

It is trivial to see how our previous description (where we used $\mathbb{C}^n$) is contained in our new formulation. Let V be an inner product space corresponding to a quantum system. Let $\chi = \{|\chi_k\rangle\,; k = 1, 2, \dots, n\}$ be an orthonormal basis in V that we associate with an observable $\hat{\chi}$ of the quantum system. An arbitrary state $|\psi\rangle$ is then described by the normalized vector

$$|\psi\rangle = \sum_k |\chi_k\rangle\langle\chi_k|\psi\rangle$$

The representation of the state vector $|\psi\rangle$ in the χ basis is given by

$$[\psi]^\chi = \begin{bmatrix} \langle\chi_1|\psi\rangle \\ \langle\chi_2|\psi\rangle \\ \\ \langle\chi_n|\psi\rangle \end{bmatrix}$$

where the coefficients $\langle\chi_k|\psi\rangle$ are given by the scalar product of $|\psi\rangle$ and $|\chi_k\rangle$ because of the orthonormality of the basis χ. By our agreed interpretation, the scalar products $\langle\chi_k|\psi\rangle$ are the probability amplitudes for the events $|\psi\rangle \to |\chi_k\rangle$. Thus the representation $[\psi]^\chi$ is nothing but what we had earlier called the state $|\psi\rangle$ in the χ space. Similarly, representations of $|\psi\rangle$ in other orthonormal bases will give the state $|\psi\rangle$ in other spaces. Thus our description neatly brings out the fact that states can be *equivalently* described in different spaces (characterized by the spectra of different observables).

Before we end this chapter, let us quickly review how the description of the quantum state evolved. We started out using a notion of a quantum state that

---

do not formally allow us to make any prediction about the *values* of the measurement outcomes.

[34]This postulate encodes the fact that *every measurement prepares a state*. Note that in the last chapter this fact was already integrated into the definition of the state that emerged from the law of definite probabilities. But in the formal description, it requires an explicit mention (which we did not do when we used $\mathbb{C}^n$). Incidentally, there is a standard jargon for this postulate: it is called the *collapse postulate* or the *postulate of state collapse*. We shall, however, introduce this term later when we discuss it in a more general setting.

was characterized by the *last measurement outcome*. Then, equipped with the superposition principle (and exchange symmetry), we were able to develop a definition of quantum state as a *collection of probability amplitudes* of events that have as a final state all possible outcomes of some arbitrary but specific observable. This definition was subsequently generalized to associate a quantum state with any set of complex numbers having some fixed cardinality (determined by the size of the spectra of the observables of the system), the squares of whose moduli add up to one. This was embodied in the statement that every *normalized vector in* $\mathbb{C}^n$ is a quantum state (in a space associated with some arbitrarily chosen observable). Finally, a (representation independent) state was defined to be a *normalized vector in an abstract inner product space*.

Problems

1. Let $\{|\chi_1\rangle, |\chi_2\rangle\}$ and $\{|\xi_1\rangle, |\xi_2\rangle\}$ be the orthonormal bases associated with the observables $\hat{\chi}$ and $\hat{\xi}$ having the respective spectra $\{\chi_1, \chi_2\}$ and $\{\xi_1, \xi_2\}$. The bases are related by

$$|\xi_1\rangle = \frac{\sqrt{3}}{2}|\chi_1\rangle + \frac{1}{2}|\chi_2\rangle \qquad \text{and} \qquad |\xi_2\rangle = -\frac{1}{2}|\chi_1\rangle + \frac{\sqrt{3}}{2}|\chi_2\rangle$$

   (a) A measurement of $\hat{\chi}$ on an initial state $|\psi\rangle$ yields χ_1. If the probability of the event in known to be $1/5$, can you say what was the initial state $|\psi\rangle$? If you knew that the amplitude for the event $|\psi\rangle \to |\chi_1\rangle$ is $1/\sqrt{5}$, would your answer be different? What is the final state after the $\hat{\chi}$ measurement?

   (b) If an instantaneous measurement of $\hat{\xi}$ is now performed, what is the probability that the outcome is ξ_2, and what is the final state when the measurement produces this outcome?

   (c) Immediately after the $\hat{\xi}$ measurement if $\hat{\chi}$ is measured again, what are the probabilities of the different possibilities? What would be your answer if you did not know the outcome of the $\hat{\xi}$ measurement[35].

   (d) By a sequence of instantaneously subsequent measurements of $\hat{\chi}$

---

[35] What you will have after the $\hat{\xi}$ measurement is known as a *statistical mixture* of states. In this case it will be a statistical mixture of the states $|\xi_1\rangle$ and $|\xi_2\rangle$ with probabilities determined by the state $|\chi_1\rangle$.

and $\hat{\xi}$ check whether it is possible to produce the states

$$(i) \qquad |\phi\rangle = \frac{1}{5}\left(4\,|\xi_1\rangle - 3\,|\xi_2\rangle\right)$$

$$(ii) \qquad |\psi\rangle = \frac{1}{10}\left(\left(3\sqrt{3} - 4\right)|\chi_1\rangle + \left(3 + 4\sqrt{3}\right)|\chi_2\rangle\right)$$

(e) Determine the probability of the event $|\psi\rangle \rightarrow |\phi\rangle$? Can the states correspond to two distinct outcomes of the same observable?

Chapter 4

Observables as Operators

In the last chapter, we started working towards a formal description of QM. The entire chapter was devoted to establishing a connection between QM and an algebraic system called inner product space. The formulation that was laid out was, however, only half baked since we were not able to assign any concrete mathematical object to quantum observables. This is what we intend to do now. In this chapter, we shall introduce a new ingredient in our language called a *linear operator*. A linear operator will turn out to be the appropriate mathematical entity to describe observables of QM. We shall see that this will allow us to rephrase the fundamental questions of QM in a very economical way, and make it easier for us to look for a road that would lead to their answers.

Linear Operators

We define an *operator* acting on a vector space as a function that associates to each member of the vector space some unique element of that vector space[1] (e.g., a $n \times n$ square matrix can be considered to be an operator acting on the vectors of $\mathbb{C}^n$ through the operation of matrix multiplication).

We shall denote operators by placing a hat on them (e.g., $\hat{A}, \hat{B}, ..., $ etc.), and their action on an arbitrary vector $|\psi\rangle$ will be written as

$$\hat{A}|\psi\rangle = |\psi\rangle'$$

We shall sometimes call $|\psi\rangle'$ the **$\hat{A}$-transform** of $|\psi\rangle$.

Now, a *linear operator* is an operator whose action preserves linearity:

$$\hat{A}[c_1|\psi_1\rangle + c_2|\psi_2\rangle] = c_1\hat{A}|\psi_1\rangle + c_2\hat{A}|\psi_2\rangle$$

where c_1, c_2 are arbitrary scalars, and $|\psi_1\rangle, |\psi_2\rangle$ are arbitrary vectors. Thus,

---

[1]In mathematics, one uses the term operator in a more general sense that refers to mappings from one vector space to another. For our purpose, it is enough to restrict the definition to mappings from a vector space to itself.

the action of a linear operator is such that the transform of a linear combination of vectors is equal to the same linear combination of the transforms of the vectors.

Since we will always be concerned with inner product spaces, the rest of the discussion will assume that our linear operators act on inner product spaces.

An operator is well defined when its action on each and every member of the space on which it acts is specified. For linear operators, it is immediately obvious that it is sufficient to specify the action of the operator on all the elements of some chosen basis. The action on every other vector is then automatically specified by virtue of linearity:

$$
\begin{aligned}
|\psi\rangle' &= \hat{A}|\psi\rangle \\
&= \hat{A}\sum_i |\chi_i\rangle\langle\chi_i|\psi\rangle \\
&= \sum_i \hat{A}|\chi_i\rangle\langle\chi_i|\psi\rangle \\
&= \sum_{i,j} |\chi_j\rangle\left\langle\chi_j\left|\hat{A}\right|\chi_i\right\rangle\langle\chi_i|\psi\rangle
\end{aligned}
$$

The summation indices will always be assumed to run over the dimension of the vector space if not specified. Here, the action of the operator $\hat{A}$ is completely specified through the specification of its action on the elements of the basis $\chi = \{|\chi_i\rangle \, ; \, i = 1, 2, \ldots n\}$, which in turn are specified through the specification of the objects[2] $\left\langle\chi_j\left|\hat{A}\right|\chi_i\right\rangle$. Note that the inner product $\left\langle\chi_j\left|\hat{A}\right|\chi_i\right\rangle$ is the coefficient of $|\chi_j\rangle$ in the expansion of $\hat{A}|\chi_i\rangle$ in the χ basis. These objects provide a representation of linear operators.

Representations of linear operators

The objects $\left\langle\chi_j\left|\hat{A}\right|\chi_i\right\rangle$ defines a $n \times n$ matrix, where n is the number of elements in the χ basis (i.e., the dimension of the vector space). Let us denote this matrix by $\left[\hat{A}\right]^\chi$.

$$
\left[\hat{A}\right]^\chi \equiv
\begin{bmatrix}
\langle\chi_1|\hat{A}|\chi_1\rangle & \langle\chi_1|\hat{A}|\chi_2\rangle & \cdots & \cdots & \langle\chi_1|\hat{A}|\chi_n\rangle \\
\langle\chi_2|\hat{A}|\chi_1\rangle & \langle\chi_2|\hat{A}|\chi_2\rangle & \cdots & \cdots & \langle\chi_2|\hat{A}|\chi_n\rangle \\
\cdots & \cdots & \cdots & \cdots & \cdots \\
\cdots & \cdots & \cdots & \cdots & \cdots \\
\langle\chi_n|\hat{A}|\chi_1\rangle & \langle\chi_n|\hat{A}|\chi_2\rangle & \cdots & \cdots & \langle\chi_n|\hat{A}|\chi_n\rangle
\end{bmatrix}
$$

---

[2] We have chosen an orthonormal basis for illustration because that is what we shall use almost all the time. It should be clear that the argument does not depend on this choice.

The equation

$$|\psi\rangle' = \sum_{i,j} |\chi_j\rangle \langle \chi_j |\hat{A}| \chi_i \rangle \langle \chi_i |\psi\rangle$$

$$\langle \chi_j |\psi\rangle' = \sum_i \langle \chi_j |\hat{A}| \chi_i \rangle \langle \chi_i |\psi\rangle$$

So we see that the matrix $\left[\hat{A}\right]^{\chi}$ transforms the χ representation $[\psi]^{\chi}$ of $|\psi\rangle$ to the χ representation $[\psi']^{\chi}$ of $|\psi\rangle'$:

$$[\psi']^{\chi} = \left[\hat{A}\right]^{\chi} [\psi]^{\chi}$$

It is therefore natural to call the matrix $\left[\hat{A}\right]^{\chi}$ the χ *representation of $\hat{A}$*, that is, the representation of the operator $\hat{A}$ in the χ space[3].

Change of representation

To obtain the representation of $\hat{A}$ in some other orthonormal basis, say $\xi = \{|\xi_i\rangle\,;\, i = 1, 2, \ldots n\}$, let us multiply the last equation of the preceding subsection by the transformation matrix $T_{(\xi,\chi)}$ that we introduced in the last chapter:

$$T_{(\xi,\chi)} [\psi']^{\chi} = T_{(\xi,\chi)} \left[\hat{A}\right]^{\chi} [\psi]^{\chi}$$

$$= T_{(\xi,\chi)} \left[\hat{A}\right]^{\chi} T^{\dagger}_{(\xi,\chi)} T_{(\xi,\chi)} [\psi]^{\chi}$$

$$[\psi']^{\xi} = T_{(\xi,\chi)} \left[\hat{A}\right]^{\chi} T^{\dagger}_{(\xi,\chi)} [\psi]^{\xi}$$

where we have used the unitarity of the transformation matrix $T_{(\xi,\chi)}$. Hence, the ξ representation $\left[\hat{A}\right]^{\xi}$ of $\hat{A}$ is clearly given by

$$\left[\hat{A}\right]^{\xi} = T_{(\xi,\chi)} \left[\hat{A}\right]^{\chi} T^{\dagger}_{(\xi,\chi)}$$

---

[3]Here the matrix elements $A^{(\chi)}_{ji}$ are given by $A^{(\chi)}_{ji} = \langle \chi_j |\hat{A}| \chi_i \rangle$. This is true for orthonormal bases. In the general case $A^{(\chi)}_{ji}$ is still equal to the j-th expansion coefficient of $\hat{A}|\chi_i\rangle$ in the χ basis, but it is no longer given by the inner product $\langle \chi_j |\hat{A}| \chi_i \rangle$. Please see appendix 'B' for details.

Problems

1. The set of vectors $B = \{|\alpha\rangle, |\beta\rangle, |\gamma\rangle\}$ forms a basis in some 3-dimensional vector space V. If the action of an operator $\hat{R}$ on the basis vectors are given by

   $$\hat{R}|\alpha\rangle = |\alpha\rangle + 2|\beta\rangle + 3|\gamma\rangle, \quad \hat{R}|\beta\rangle = 3|\alpha\rangle + |\beta\rangle + 2|\gamma\rangle, \quad \hat{R}|\gamma\rangle = 2|\alpha\rangle + 3|\beta\rangle + |\gamma\rangle$$

   (a) What will be the action of $\hat{R}$ on the vector: $|\psi\rangle = |\alpha\rangle - 2|\beta\rangle + |\gamma\rangle$?

   (b) Determine the matrix representation of the operator $\hat{R}$ and the representation of $\hat{R}|\psi\rangle$ in the B basis.

2. Consider an operator $\mathcal{F}$ acting on $\mathbb{C}^2$ such that

   $$\mathcal{F}(x, y)^T = (2x + 3y, 4x - 5y)^T$$

   where $(x, y)^T \in \mathbb{C}^2$. Show that $\mathcal{F}$ is a linear operator and find its representation in the standard basis $\left\{(1, 0)^T, (0, 1)^T\right\}$ and the orthonormal basis $\left\{(1/\sqrt{2})(1, 1)^T, (1/\sqrt{2})(-1, 1)^T\right\}$.

3. Let $\hat{M}$ be a linear operator acting on $\mathbb{C}^2$. If the action of $\hat{M}$ on the orthonormal basis elements $(1, 0)^T$ and $(0, 1)^T$ are given by $\hat{M}(1, 0)^T = (1, 1)^T$ and $\hat{M}(0, 1)^T = (1, -2)^T$. Find the action of $\hat{M}$ on an arbitrary element $(x, y)^T$ of $\mathbb{C}^2$. What will be the matrix representation of $\hat{M}$?

4. An operator $\hat{A}$ is defined on $\mathbb{C}^2$ in the standard basis by the matrix,

   $$\hat{A} = \begin{bmatrix} 5 & 2 \\ -1 & 4 \end{bmatrix}$$

   Determine the matrix representing $\hat{A}$ in the basis,

   $$S_1 = \left\{ \frac{1}{\sqrt{2}} \begin{bmatrix} 1 \\ 1 \end{bmatrix}, \frac{1}{\sqrt{2}} \begin{bmatrix} 1 \\ -1 \end{bmatrix} \right\}$$

5. The outer product $|\alpha\rangle\langle\beta|$ associated with an ordered pair of vectors $(|\alpha\rangle, |\beta\rangle)$ in a vector space V is an operator defined by

   $$[|\alpha\rangle\langle\beta|]|\psi\rangle = |\alpha\rangle\langle\beta|\psi\rangle$$

   where $|\psi\rangle$ is an arbitrary vector in V.

   (a) Show that the operator is linear.

   (b) If $B = \{|a\rangle, |b\rangle\}$ is an orthonormal basis, determine the representation $\left[\hat{Q}\right]^B$ where $\hat{Q}$ is defined by

   $$\hat{Q} = p|a\rangle\langle a| + q|b\rangle\langle b| + r|a\rangle\langle b| + s|b\rangle\langle a|$$

   where p, q, r, s are scalars. Here, addition of operators, and multiplication of operators by a scalar are defined in a *natural* way.

Eigenvalues and eigenvectors

Let $\hat{A}$ be a linear operator acting on a vector space V. If for some vector $|\psi\rangle$ in V and some scalar λ

$$\hat{A}\,|\psi\rangle \;=\; \lambda\,|\psi\rangle$$

then we say that $|\psi\rangle$ is an **eigenvector** of $\hat{A}$ with **eigenvalue** λ. The above equation is called the **eigenvalue equation** of $\hat{A}$. The full set of eigenvalues is called the **spectrum** of the operator.

In general, there can be more than one *linearly independent* eigenvector corresponding to a given eigenvalue. As for linearly dependent eigenvectors, it is easy to see that

if a set of eigenvectors of a linear operator have the same eigenvalue, then every linear combination of those eigenvectors is also an eigenvector of the operator with that eigenvalue.

So given a linearly independent set of eigenvectors with a specific eigenvalue, there exists a linearly dependent set of infinitely many eigenvectors with that same eigenvalue which can be constructed simply by taking the linear combinations of the given set of independent eigenvectors. An immediate corollary to the above theorem is that

the eigenvectors corresponding to different eigenvalues are always linearly independent[4].

If for a linear operator, there are two or more linearly independent eigenvectors corresponding to the same eigenvalue, then the eigenvalue is said to be **degenerate**. If an eigenvalue has a unique eigenvector up to scalar multiples (i.e., just one linearly independent eigenvector), it is called **nondegenerate**. The maximum number of linearly independent eigenvectors that an eigenvalue can have is called the **degree of degeneracy** of that eigenvalue. If an eigenvalue has maximally g linearly independent eigenvectors, then it is said to be **g-fold degenerate**. The set of all linear combinations of the g independent eigenvectors of a g-fold degenerate eigenvalue forms a subset of V, and this subset also forms a vector space. It is called the **eigensubspace** (or simply, the **eigenspace**) of the eigenvalue. Clearly, every eigenvector of an eigenvalue must belong to its eigensubspace.

---

[4]This follows directly by *contrapositivity*: for two statements A and B

$$\textbf{if } (A \implies B) \textbf{ then } ((\text{not } B) \implies (\text{not } A))$$

Determination of eigenvalues and eigenvectors

If we cast the eigenvalue equation for a linear operator $\hat{A}$ in a representation with respect to some basis, say $\chi = \{|\chi_i\rangle\,;\,i = 1, 2, \ldots, n\}$, we get a system of n linear homogeneous equations:

$$\left[\hat{A}\right]^{\chi}[\psi]^{\chi} \;=\; \lambda\,[\psi]^{\chi}$$

Writing the components of $\left[\hat{A}\right]^{\chi}$ as A_{ij}, the n equations read

$$\sum_{j=1}^{n} A_{ij}\psi^{j} \;=\; \lambda\psi^{i}$$

with i taking values from 1 to n. Here ψ^{i} and A_{ij} are given by

$$|\psi\rangle \;=\; \sum_{i=1}^{n} \psi^{i}\,|\chi_i\rangle$$

$$\text{and} \qquad \hat{A}\,|\chi_j\rangle \;=\; \sum_{i=1}^{n} A_{ij}\,|\chi_i\rangle$$

Writing $\lambda\psi^{i} = \lambda\sum_{j}\delta_{ij}\psi^{j}$, where δ_{ij} is the Kronecker delta function, we have

$$\sum_{j=1}^{n}\left(A_{ij} - \lambda\delta_{ij}\right)\psi^{j} \;=\; 0$$

Switching back to matrix notation, the above system of equations become

$$\left(\left[\hat{A}\right]^{\chi} - \lambda\,[I]\right)[\psi]^{\chi} = 0$$

where $[I]$ is the $n \times n$ identity matrix. This system of homogeneous equations can have a nontrivial solution if and only if the determinant of the matrix $([A]^{\chi} - \lambda\,[I])$ is zero:

$$\det\left([A]^{\chi} - \lambda\,[I]\right) \;=\; 0$$

This is known as the ***characteristic equation*** of the operator.

The characteristic equation is a polynomial equation in λ of degree n. It therefore has n roots (which are not necessarily distinct). Thus, the eigenvalues λ of the operator $\hat{A}$ are the roots of its characteristic equation. The solution(s) of the eigenvalue equation corresponding to a given root yields the eigenvector(s) of the corresponding eigenvalue in the chosen representation[5].

The number of times an eigenvalue (i.e., a given root of the characteristic

---

[5] Please see appendix 'B' for details.

equation) is repeated in the characteristic equation is called the **algebraic multiplicity** of the eigenvalue. The degree of degeneracy of the eigenvalue is called its **geometric multiplicity**.

The geometric multiplicity of an eigenvalue is at least one, and at most equal to its algebraic multiplicity[6].

Diagonalization

An operator is said to have been *diagonalized* if the operator is represented as a diagonal matrix. It is easy to see the following result:

The representation of an operator is diagonal (i.e., given by a diagonal matrix) if and only if the basis comprises its own eigenvectors. When this happens, the eigenvalues appear as the diagonal elements of the representation matrix.

Solving the eigenvalue problem to determine a set of eigenvectors that can constitute a basis is, therefore, often referred to as *diagonalization*. Of course, *it is not guaranteed that such a basis will always exist.* Clearly,

one can construct a basis out of the eigenvectors of an operator if and only if, for all eigenvalues of the operator, the geometric multiplicity is equal to the algebraic multiplicity.

Such a linear operator is said to be *diagonalizable*.

Hermitian operators

Let us for this subsection, use the following notation for inner products:

$$\langle \phi | \psi \rangle \equiv (|\phi\rangle, |\psi\rangle)$$

If for a linear operator $\hat{A}$ acting on some inner product space V, there exists an operator $\hat{A}^\dagger$ such that

$$\left(|\phi\rangle, \hat{A} |\psi\rangle \right) = \left(\hat{A}^\dagger |\phi\rangle, |\psi\rangle \right)$$

for all vectors $|\phi\rangle, |\psi\rangle$ in V, then the operator $\hat{A}^\dagger$ is called the **Hermitian conjugate** to the operator $\hat{A}$. The Hermitian conjugate of an operator is also called the **adjoint** of the operator.

A **Hermitian** or **self adjoint operator** is defined by the condition[7]

$$\hat{A}^\dagger = \hat{A}$$

---

[6] For a proof of this important result, please see appendix 'B'.

[7] Actually, a self adjoint operator is not in general the same as a Hermitian operator, but on a finite dimensional space they are identical.

Thus, for a Hermitian operator $\hat{A}$ we have

$$\left(|\phi\rangle, \hat{A}|\psi\rangle\right) = \left(\hat{A}|\phi\rangle, |\psi\rangle\right)$$

for all vectors $|\phi\rangle, |\psi\rangle$ in V.

Incidentally, an **anti-Hermitian operator** $\hat{A}$ is defined by the condition

$$\hat{A}^\dagger = -\hat{A}$$

Important properties of Hermitian operators

The following properties of Hermitian operators will turn out to be important in QM[8].

- *The eigenvalues of a Hermitian operator are real.*

- *The eigenvectors of a Hermitian operator corresponding to different eigenvalues are orthogonal.*

- *An orthonormal basis can always be constructed out of the eigenvectors of any Hermitian operator acting on a finite dimensional space (this also means that a Hermitian operator is diagonalizable by a unitary transformation).*

Conversely

- *If the eigenvalues of a linear operator are real, and if an orthonormal basis can be constructed out of its eigenvectors, then the operator is Hermitian.*

For finite dimensional spaces the last property, clearly, provides an *equivalent characterization of Hermiticity*.

Incidentally, Hermitian operators that act on infinite dimensional spaces may not have a complete set of eigenvectors[9]. In our discussion, Hermitian operators will, however, always refer to operators that have a complete set of eigenvectors.

Representation of Hermitian operators

In finite dimensional spaces, the representation of the Hermitian conjugate of an operator $\hat{A}$ in an orthonormal basis can simply be obtained by taking the Hermitian conjugate of the matrix representing $\hat{A}$. That is

$$\left[\hat{A}^\dagger\right]^\chi = \left(\left[\hat{A}\right]^\chi\right)^\dagger$$

---

[8]The proofs have been provided in appendix 'B'. However, it is better if you attempt to prove the results yourself before you look them up.

[9]Hermitian operators which do have a complete set of eigenvectors have been given a special name. They are called *observables* (inspired by their use in QM as we shall see).

Clearly, the representation of a Hermitian operator in an orthonormal basis is a Hermitian matrix.

Algebra of operators

We define ***addition*** $\hat{A} + \hat{B}$***, of two operators*** $\hat{A}$ and $\hat{B}$, and the ***multiplication*** $c\hat{A}$ ***of an operator*** $\hat{A}$ ***by a scalar*** c, respectively, by the following rules:

$$\left(\hat{A} + \hat{B}\right) |\psi\rangle = \hat{A} |\psi\rangle + \hat{B} |\psi\rangle$$

$$\left(c\hat{A}\right) |\psi\rangle = c\left(\hat{A} |\psi\rangle\right)$$

where $|\psi\rangle$ is an arbitrary vector in the vector space. The ***multiplication*** $\hat{A}\hat{B}$ ***of two operators*** $\hat{A}$ ***and*** $\hat{B}$ is defined by *successive action* on $|\psi\rangle$:

$$\left(\hat{A}\hat{B}\right) |\psi\rangle = \hat{A}\left(\hat{B} |\psi\rangle\right)$$

Note that multiplication of operators do not in general commute. That is, in general $\hat{A}\hat{B} \neq \hat{B}\hat{A}$. It is clear that algebraic combinations of operators, such as polynomials, can now be naturally defined on linear operators.

Representations of algebraic combinations of operators

It is not hard to see that the representation of the sum of two operators is given by the sum of the representations. Similarly, the representation of the scalar multiple or product of linear operators are given, respectively, by the scalar multiple and the product of the representations:

$$\left[\hat{A} + \hat{B}\right]^{\chi} = \left[\hat{A}\right]^{\chi} + \left[\hat{B}\right]^{\chi}$$

$$\left[c\hat{A}\right]^{\chi} = c\left[\hat{A}\right]^{\chi}$$

$$\left[\hat{A}\hat{B}\right]^{\chi} = \left[\hat{A}\right]^{\chi}\left[\hat{B}\right]^{\chi}$$

where $\hat{A}$ and $\hat{B}$ are linear operators acting on some space V, and c is a scalar. As always, χ denotes an orthonormal basis in V where the operators have been represented.

Problems

1. Find out the eigenvalues and normalized eigenvectors of the Hermitian operator

$$\hat{X} = \begin{bmatrix} 2 & 3 - 4i \\ 3 + 4i & 2 \end{bmatrix}$$

acting on $\mathbb{C}^2$. Show explicitly that the eigenvectors corresponding to different eigenvalues are orthogonal. Construct the transformation matrix T using the normalized eigenvectors that would diagonalize $\hat{X}$.

2. A 2×2 Hermitian operator $\hat{S}$ acting on $\mathbb{C}^2$ has an eigenvalue σ with algebraic multiplicity 2. Write down the representation of $\hat{S}$ in the basis $\left\{ (a,b)^T, (c,d)^T \right\}$. Can you see (without attempting to construct the eigenvectors) why the following matrix cannot be diagonalized?

$$\begin{bmatrix} 1 & 0 \\ -2 & 1 \end{bmatrix}$$

3. The eigenvalues of the following matrices

$$\begin{bmatrix} 4 & -2 & 3 \\ 1 & 1 & 3 \\ 1 & -2 & 6 \end{bmatrix} \quad \text{and} \quad \begin{bmatrix} -2 & 0 & 1 \\ 1 & 1 & 0 \\ 0 & 0 & -2 \end{bmatrix}$$

are $(3, 5)$ and $(1, -2)$ respectively. Find out their algebraic and geometric multiplicities, and hence determine their diagonalizability. Evaluate the possibility of diagonalizing them by a unitary transformation matrix (if diagonalizable).

4. For each of the Hermitian matrices given below, construct orthonormal bases comprising their eigenvectors

$$i) \begin{bmatrix} 2 & -1 & 0 \\ -1 & 2 & -1 \\ 0 & -1 & 2 \end{bmatrix} \qquad ii) \begin{bmatrix} 0 & 1 & 1 \\ 1 & 0 & 1 \\ 1 & 1 & 0 \end{bmatrix}$$

Are these bases unique?

5. Solve the eigenvalue problems for the Hermitian operators

$$i) \begin{bmatrix} 0 & 1 \\ 1 & 0 \end{bmatrix} \qquad ii) \begin{bmatrix} 0 & -i \\ i & 0 \end{bmatrix}$$

Using the solutions, directly write down the eigenvalues and eigenvectors of the matrices

$$i) \begin{bmatrix} 0 & 0 & 1 \\ 0 & 7 & 0 \\ 1 & 0 & 0 \end{bmatrix} \qquad ii) \begin{bmatrix} 3 & 0 & 0 \\ 0 & 1 & 2i \\ 0 & -2i & 1 \end{bmatrix}$$

(Hint: Try to identify the matrices you have solved for in the above matrices. Change the ordering of the basis vectors, if required, and exploit the block-diagonal structure of the matrices.)

6. Let $\hat{Q}$ be an operator with an eigenvalue q and let the corresponding eigenvector be $|q\rangle$. Show that the operator $f\left(\hat{Q}\right)$ will also have $|q\rangle$ as an eigenvector with $f(q)$ as the corresponding eigenvalue where f is a polynomial function.

7. Show that the *projection operator* $\hat{P}_i$ associated with the eigenspace of a g_i-fold degenerate eigenvalue A_i of an observable $\hat{A}$ can be written as a linear combination of outer products (please see the section on *"Direct Sum of Subspaces"* in appendix 'B' for the definition of *projection operator*):

$$\hat{P}_i = \sum_{r=1}^{g_i} |A_i^r\rangle \langle A_i^r|$$

where $|A_i^r\rangle$ comprise a set of orthonormal eigenvectors with eigenvalue A_i. Hence, demonstrate that the observable $\hat{A}$ can be represented as

$$\hat{A} = \sum_i A_i \hat{P}_i$$

where i runs over all the eigenvalues of $\hat{A}$.

8. Prove that the set Θ of all linear operators acting on some vector space, itself forms a vector space under the functions, addition and multiplication by a scalar.

9. Show that, if $\hat{A}, \hat{B}, \hat{C}$ are linear operators acting on some vector space, and c is a scalar, then the following properties hold[10]:

   (a) $\hat{A}\left(\hat{B} + \hat{C}\right) = \hat{A}\hat{B} + \hat{A}\hat{C}$

   (b) $\left(\hat{A} + \hat{B}\right)\hat{C} = \hat{A}\hat{C} + \hat{B}\hat{C}$

   (c) $c\left(\hat{A}\hat{B}\right) = \left(c\hat{A}\right)\hat{B} = \hat{A}\left(c\hat{B}\right)$

   (d) $\left(\hat{A}\hat{B}\right)\hat{C} = \hat{A}\left(\hat{B}\hat{C}\right)$

10. Prove the following properties on Hermitian conjugation: If $\hat{A}, \hat{B}$ are linear operators acting on an inner product space V, c is a scalar, and $|\psi\rangle, |\phi\rangle$ are arbitrary vectors in V, then

   (a) $\left(\hat{A} + \hat{B}\right)^{\dagger} = \hat{A}^{\dagger} + \hat{B}^{\dagger}$

   (b) $\left(c\hat{A}\right)^{\dagger} = c^* \hat{A}^{\dagger}$

---

[10]A vector space over $\mathbb{C}$, on which multiplication is defined satisfying these properties is called an *associative algebra,* or more precisely, an associative algebra over the field $\mathbb{C}$. If the last property is dropped, it becomes simply an *algebra* over the field $\mathbb{C}$.

(c) $\left(\hat{A}\hat{B}\right)^{\dagger} = \hat{B}^{\dagger}\hat{A}^{\dagger}$

(d) $\left\langle \psi|\hat{A}|\phi \right\rangle^{*} = \left\langle \phi|\hat{A}^{\dagger}|\psi \right\rangle$

Describing Observables by Linear Operators

Every observable in QM has associated with it a spectrum whose members are real scalars. Again, according to the second postulate of QM, every member of the spectrum of an observable is associated with a state vector, and the set of state vectors associated with the full spectrum of an observable forms an orthonormal basis. Now, to define a linear operator, we have seen that it is sufficient to specify the action of the operator on every member of some basis. One can then, trivially, associate with every observable in QM a unique linear operator defined by the following simple rule:

If $\hat{A}$ is an observable having the spectrum $\{A_i; i = 1, 2, \ldots, n\}$, then we define a linear operator $\hat{A}$ (which we choose to denote by the same symbol as that of the observable) by

$$\hat{A}|A_i\rangle = A_i|A_i\rangle$$

where the vectors $|A_i\rangle$ are the state vectors corresponding to the members A_i of the spectrum of the observable $\hat{A}$.

By construction, the spectrum of the observable $\hat{A}$ then becomes the set of *nondegenerate*, real eigenvalues of the operator $\hat{A}$, and the state vectors corresponding to the members of the spectrum of the observable $\hat{A}$ become a complete set of orthonormal eigenvectors of the operator $\hat{A}$ (often referred to as **eigenstates** of the observable). Now, we know that an operator with real eigenvalues and a basis of orthonormal eigenvectors is Hermitian. Thus, we can make the following assertion (see Figure 4.1):

- *Every observable in QM can be associated with a linear, Hermitian operator having only nondegenerate eigenvalues*[11]. *The possible outcomes of measurement of the observable are the eigenvalues of the operator, and the states associated with the outcomes are the normalized eigenvectors of the respective eigenvalues*[12].

---

[11]The restriction to nondegenerate eigenvalues will be lifted when we discuss more general scenarios.

[12]The spectrum of the observable is thus the eigenvalue spectrum of the corresponding operator.

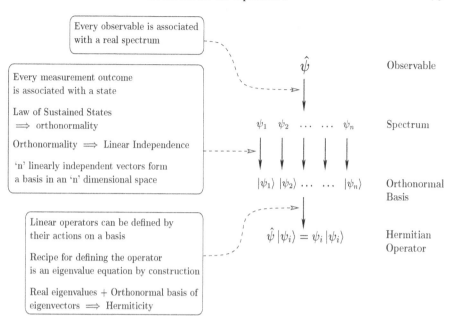

FIGURE 4.1: Construction of the linear Hermitian operator associated with an observable.

We can generalize this observation to assert that the converse to this statement is also true.

- *Every linear, Hermitian operator having only nondegenerate eigenvalues can be imagined to be associated with an observable in QM. The eigenvalues of the operator would correspond to the possible outcomes of measurement of the observable, and the normalized eigenvectors of the eigenvalues would correspond to the states associated with respective outcomes[13].*

This is possible because of the following reasons. Firstly, the eigenvalues of a Hermitian operator are real, so that one can interpret them as measurement outcomes. Secondly, if the eigenvalues are nondegenerate, the eigenvectors (up to a multiplicative scalar) can be labelled by a unique eigenvalue. Finally, for a Hermitian operator with nondegenerate eigenvalues, collecting the normalized eigenvectors for each eigenvalue will give us an orthonormal basis[14].

---

[13]This generalization is similar in spirit to what we did when we made the assertion that *every* normalized vector in $\mathbb{C}^n$ is a quantum state. This was a generalization based on the observation that quantum states, according to the previous definition, were normalized vectors in $\mathbb{C}^n$.

[14]Here again, let us point out that it is irrelevant how one can devise an experiment that *measures* the observable corresponding to an arbitrary Hermitian operator. The important point is that we are obliged to admit that it exists.

Postulates of Quantum Mechanics - Version 2

By virtue of the association of quantum observables and Hermitian operators established in the foregoing section, we will now replace the second postulate of QM by two assertions which will characterize a quantum observable more completely. Let us incorporate this change, and write down the postulates of QM once again.

1. Every system is associated with an inner product space. States of the system are the normalized vectors belonging to this space.

2. Every observable of the system is associated with a linear Hermitian operator with nondegenerate eigenvalues acting on the state space.

3. The eigenvalues of the Hermitian operator associated with an observable constitute the measurement spectrum of the observable.

4. If a measurement of an observable $\hat{\phi}$ is made that yields an eigenvalue ϕ_i, soon after the measurement the new state becomes the normalized eigenstate $|\phi_i\rangle$ corresponding to the eigenvalue ϕ_i.

5. The probability $P[\psi_i \to \phi_j]$ of the event $\psi_i \to \phi_j$ of getting an outcome ϕ_j upon measurement of an observable $\hat{\phi}$ on a state $|\psi_i\rangle$ is given by the *Born rule*:

$$P[\psi_i \to \phi_j] \quad = \quad |\langle \phi_j | \psi_i \rangle|^2$$

where the scalar product $\langle \phi_j | \psi_i \rangle$ is called the probability amplitude of the event $\psi_i \to \phi_j$.

Note that by postulating an observable to be a linear Hermitian operator, in one stroke, not only have we furnished a complete mathematical characterization of a quantum observable, we have rendered it completely unnecessary to require that to every outcome of an observable there corresponds a quantum state, and the set of all such states form an orthonormal basis. Eigenvalues are naturally associated with eigenvectors which can be taken to be normalized. This association is one-to-one (at least, up to a global phase) if eigenvalues are assumed to be nondegenerate. Finally, Hermiticity ensures that the set of normalized eigenvectors associated with the full set of eigenvalues will form an orthonormal basis.

Quantization

Rephrasing the target questions

So far, we have been able to connect quantum systems with inner product spaces, quantum states with normalized vectors, and quantum observables with linear Hermitian operators, so that the laws of QM manifest themselves as concrete statements involving these mathematical constructs. But the curious mind would surely wonder, what does it really buy us[15]? Are there any practical benefits that we can reap out of this exercise in trying to find the answers to the target questions of QM? The answer is "yes", and this is what we wish to discuss now.

Suppose, we are concerned with the measurement of the observables $\hat{\alpha}$, $\hat{\beta}$, $\hat{\gamma}$, ..., etc., of some system. Assuming that we have chosen to work in the χ space, the respective representations of the observables $\hat{\alpha}$, $\hat{\beta}$, $\hat{\gamma}$ ... would be the matrices $[\hat{\alpha}]^{\chi}$, $\left[\hat{\beta}\right]^{\chi}$, $[\hat{\gamma}]^{\chi}$, ..., etc. If we solve the eigenvalue equations for these matrices, the eigenvalues will give the spectra $\{\alpha_i\}, \{\beta_j\}, \{\gamma_k\}$... of the observables, and the probability of an event such as $\alpha_i \to \beta_j$ will be given by the scalar product of the normalized eigenvectors $[\alpha_i]^{\chi}$ and $[\beta_j]^{\chi}$:

$$
\begin{aligned}
P[\alpha_i \to \beta_j] &= |\langle \beta_j | \alpha_i \rangle|^2 \\
&= \left| ([\beta_j]^{\chi})^{\dagger} [\alpha_i]^{\chi} \right|^2
\end{aligned}
$$

The prescription for answering the target questions of QM is then to solve the eigenvalue equations for the Hermitian operators associated with the observables in question in some representation. So, with the "*observable – linear Hermitian operator*" association, the fundamental problem of QM (for instantaneously subsequent measurements) has been translated to an eigenvalue problem[16]:

Instead of asking
> *what are the possibilities and what are the probabilities?*

we can ask
> *what are the eigenvalues and what are the eigenvectors?*

---

[15] Of course, there are the advantages of having a formal description, as we have discussed in the first chapter. But we are looking for more dividends here.

[16] It is not always the case that in a theory the target questions come with a clue that leads to a strategy for its solution. In this situation, quite often, one tries to rephrase the question in a way that might offer a new perspective leading to an idea for a method of solution. This is what we are actually doing here.

Quantization rules

With the target questions of QM rephrased, it is evident that there are two steps involved in laying down a quantum mechanical description for a system. The first step is to specify the state space: the inner product space in which the states of the system live, and on which the observables of the system act. The second step is to identify the observables of interest, and specify their explicit representations in some basis. The eigenvalue problems for the observables can, of course, only be solved if the representations for the observables are explicitly known in some basis. This whole information is, naturally, system specific, and has to be supplied as a separate set of assumptions that must go into the quantum algorithm as an input. These system specific assumptions are called **quantization rules**. Specifying the quantization rules is often referred to as *quantizing* the system. Specifically, quantization rules furnish the following information:

1. The state space of the system.

2. The representations of all observables of interest in some basis[17].

The way in which this information is encoded in the quantization rules is, unfortunately, not always very straightforward. In practice, for most real life systems, one uses an *intermediate classical system* to lay down the recipe for quantization. The phrase "quantizing the system" actually originates in this context. Having an intermediate classical system is, however, not a logical requirement for defining quantization. It is only a *way of guessing* that has proved to be very successful in an overwhelmingly large number of cases. We will discuss quantization of systems that use the scaffolding of intermediate classical systems later on. For now, we wish to discuss quantization without making any reference to intermediate classical systems.

In the simplest case, the inner product space corresponding to a system is specified by specifying an orthonormal basis[18]. The orthonormal basis is specified by specifying the spectrum of some observable. Each element of the basis is labelled by a unique member of the spectrum of the observable. This spectrum, incidentally, will often be called the **primitive spectrum**, and the corresponding basis will often be referred to as the **first basis** in this book.

Now, let us come to the representation of observables. It is trivial to conceive the matrix representation of the observable that furnishes the primitive spectrum which defines the first basis. The primitive spectrum comprises the set of eigenvalues of the observable and the basis elements are its eigenvectors. But we know that the representation of an observable in its own space (i.e., in a basis made out of its own eigenvectors) is simply a diagonal matrix

---

[17]If we could determine the eigenvalues and eigenvectors without specifying the representations, that would also be acceptable but that will not generally be the case.

[18]For this purpose, the orthonormality of the basis is actually not essential. If we specify the inner products of the basis elements, it is good enough. However, more often than not one tends to use an orthonormal basis.

with the eigenvalues as the diagonal elements. For example, the matrix $[\hat{\chi}]^\chi$ corresponding to the observable $\hat{\chi}$ in its own space is given by

$$\langle \chi_i | \hat{\chi} | \chi_j \rangle = \chi_i \delta_{ij}$$

It is clear that a knowledge of this representation is no more (or no less) than a knowledge of the primitive spectrum. The representations of observables other than the one which furnishes the primitive spectrum comprise the most nontrivial part of the quantization rules. Let us illustrate this using a simple toy example.

A toy example of quantization[19]

We consider a quantum system which is defined by the following quantization rules:

1. The system has three *basic* observables, $\hat{A}$, $\hat{B}$ and $\hat{C}$. We call these observables "basic" because we assume that every other observable for this system can be expressed as a function of these basic operators.

2. The observable $\hat{C}$ has two possible outcomes, which we *label* as $+\kappa$ and $-\kappa$, where κ is some (possibly dimensionful) positive real number. The spectrum $\{+\kappa, -\kappa\}$ of $\hat{C}$ is our primitive spectrum.

   This furnishes a first basis: $\{|+\kappa\rangle, |-\kappa\rangle\}$. The representation of the basis vectors in their own space are trivially given by

   $$[+\kappa]^C = \begin{bmatrix} 1 \\ 0 \end{bmatrix} \qquad \text{and} \qquad [-\kappa]^C = \begin{bmatrix} 0 \\ 1 \end{bmatrix}$$

   The representation of $\hat{C}$ in this basis (i.e., in its own space) is also trivial

   $$\left[\hat{C}\right]^C = \kappa \begin{bmatrix} 1 & 0 \\ 0 & -1 \end{bmatrix}$$

3. The representations of the operators $\hat{A}$ and $\hat{B}$ in the C space are *assumed* to be given by

   $$\left[\hat{A}\right]^C = \kappa \begin{bmatrix} 0 & 1 \\ 1 & 0 \end{bmatrix} \qquad \text{and} \qquad \left[\hat{B}\right]^C = \kappa \begin{bmatrix} 0 & -i \\ i & 0 \end{bmatrix}$$

These assumptions are all we need to make all possible predictions (for instantaneously subsequent measurements) about this system[20].

---

[19] Incidentally, this is an example of a 2-level system: a system whose states live in an inner product space that is 2-dimensional.

[20] Please work out problem '3' in the next problem set.

I am sure you are wondering, where on earth did the representations $\left[\hat{A}\right]^C$ and $\left[\hat{B}\right]^C$ come from? Well, that is precisely the point I wanted to make. These representations are nontrivial assumptions, and there is no general prescription to guess them any more than there is a prescription to guess a Hooke's law of elasticity, or a Newton's law of gravitation. The purpose of this toy example was to point out what sort of mathematical input is required to carry out the programme in QM. That is why we did not even care to mention, what kind of physical system our toy example describes, and what the observables $\hat{A}$, $\hat{B}$ and $\hat{C}$ mean physically. In real life, the nontrivial representations have to be obtained through educated guesswork. Later on, we shall describe some of the schemes for guessing representations that have worked in a large number of actual physical problems. For now, let us be content with these god given representations to see what we can do with them.

Expectation Values

Since QM is essentially statistical in nature, it makes sense to talk about the mean value of an observable. This refers to the average value in the outcome when measurement of a specified observable is made on a large number of identically prepared systems (i.e., on systems that are in the same initial state)[21]. In fact, it turns out that the quantity of interest to the experimentalist is, quite often, the average value of an observable. It is, for various practical reasons, the quantity that is actually accessible to experiments.

Now, an average value of a statistical quantity naturally depends on the size of the sample on which the experiment is carried out. In our case, the sample size means the number of systems prepared in an identical quantum state on which measurement of the observable in question is performed. As far as quantum *prediction* is concerned, the relevant quantity would be the *idealized average* defined as the limiting value of the average as the sample size is taken to infinity. Such a quantity is called the ***expectation value*** of the observable.

If $\bar{A}_N$ is the average value for an observable $\hat{A}$ (having m possible outcomes) when the measurement is performed on N identical systems, then

$$\bar{A}_N = \frac{1}{N} \sum_{i=1}^{m} N_i A_i$$

where N_i is the number of systems that has yielded the outcome A_i. Clearly,

---

[21]Such a collection of systems is called an *ensemble*. One says that the measurement is performed on an ensemble of identically prepared systems.

$\sum_{i=1}^{m} N_i = N$. The expectation value $\langle A \rangle$ of $\hat{A}$ is then defined by

$$\langle A \rangle = \operatorname*{Lim}_{N \to \infty} \bar{A}_N$$

Let us derive an expression for the expectation value of the observable $\hat{A}$ if the system is in an initial state $|\psi\rangle$. We have

$$\operatorname*{Lim}_{N \to \infty} \bar{A}_N = \operatorname*{Lim}_{N \to \infty} \left(\frac{1}{N} \right) \sum_{i=1}^{m} N_i A_i$$

$$= \sum_{i=1}^{m} \operatorname*{Lim}_{N \to \infty} \left(\frac{N_i}{N} \right) A_i$$

$$= \sum_{i=1}^{m} P_i A_i$$

where $P_i = \operatorname*{Lim}_{N \to \infty} (N_i/N)$ is, by definition, the probability for the outcome A_i. Hence

$$\langle A \rangle = \sum_{i=1}^{m} P_i A_i$$

But, if the system is in a state $|\psi\rangle$, then $P_i = |\langle A_i|\psi\rangle|^2$. So

$$\langle A \rangle = \sum_{i=1}^{m} \langle \psi|A_i \rangle \langle A_i|\psi \rangle A_i$$

$$= \sum_{i=1}^{m} \left\langle \psi|\hat{A}|A_i \right\rangle \langle A_i|\psi \rangle$$

$$= \langle \psi| \hat{A} \sum_{i=1}^{m} |A_i \rangle \langle A_i|\psi \rangle$$

where we have used the relations: $\langle A_i|\psi \rangle^* = \langle \psi|A_i \rangle$, $\hat{A}|A_i \rangle = A_i|A_i \rangle$ and $|\psi\rangle = \sum_{i=1}^{m} |A_i \rangle \langle A_i|\psi \rangle$. Using the last formula again, we finally have

$$\langle A \rangle = \left\langle \psi|\hat{A}|\psi \right\rangle$$

Problems[22]

1. Which of the following operators will describe a quantum observable?

$$i)\ \left[\hat{A}\right] = \begin{bmatrix} 1 & 1 & -1 \\ -2 & 3 & 0 \\ -2 & 1 & 2 \end{bmatrix} \qquad ii)\ \left[\hat{B}\right] = \begin{bmatrix} 3 & 2-i & -3i \\ 2+i & 0 & 1-i \\ 3i & 1+i & 0 \end{bmatrix}$$

2. Two observables $\hat{U}$ and $\hat{V}$ are represented as

$$\left[\hat{U}\right] = \mu \begin{bmatrix} 11 & 0 \\ 0 & 19 \end{bmatrix} \qquad \text{and} \qquad \left[\hat{V}\right] = \frac{\lambda}{2} \begin{bmatrix} 1 & 1 \\ 1 & 1 \end{bmatrix}$$

where μ and λ are dimensionful constants. If a measurement of $\hat{U}$ that yields an outcome 19μ is followed up by an instantaneously subsequent measurement of $\hat{V}$, what are the possible outcomes and what are their probabilities? Do not compute any inner product.

3. Consider the toy example of a quantized system which was discussed in the text.

   (a) Show that any observable (i.e., an arbitrary Hermitian operator) can indeed be expressed as a function of the operators $\hat{A}$, $\hat{B}$ and $\hat{C}$.

   (Hint: Observe that $\hat{A}^2 = \hat{B}^2 = \hat{C}^2 = \kappa^2 \hat{I}_2$ where $\hat{I}_2$ is the 2-dimensional identity matrix. Now write out the general form of a complex, 2-dimensional Hermitian matrix, and show that it can be expressed as a linear combination of the matrices $\hat{A}$, $\hat{B}$, $\hat{C}$ and $\hat{I}_2$.)

   (b) If a measurement of the observable

   $$\hat{M} = \frac{3\omega}{\kappa}\hat{A} + \frac{4\omega}{\kappa}\hat{B} + \frac{2\omega}{\kappa^2}\hat{C}^2$$

   is made on the system which is in a state $[\psi]^C = (1/\sqrt{2})\,(1, -1)^T$, find out the possible outcomes of the measurement and the probabilities of the different possibilities. Here ω is another dimensionful constant.

   (c) A sequence of four instantaneously subsequent measurements $\hat{A}, \hat{B}, \hat{A}$ and $\hat{B}$ are performed on the system. If the outcome of the first measurement is A_1, find out the probability that the outcome of the last measurement is B_2. Here, by convention, the spectrum of an observable $\hat{X}$ is written as $\{X_1, X_2\}$ with $X_1 > X_2$.

---

[22] None of the problems in this problem set requires an explicit solution of an eigenvalue problem. You can use the work that you have done in the last problem set.

(d) An observable $\hat{N}$ is represented in the C space by the Hermitian matrix

$$\left[\hat{N}\right]^C = \kappa \begin{bmatrix} 2 & -i \\ i & 2 \end{bmatrix}$$

From the results of your previous problems (i.e., without solving any more eigenvalue equations), write down the spectrum of this observable, and determine the probabilities of the events (i) $N_1 \to B_1$ and (ii) $N_1 \to A_2$ (where the same convention for writing the spectral elements have been used as in part 'c').

4. An observable is described by an operator $\hat{P}$, which is represented in some basis as

$$\left[\hat{P}\right] = \begin{bmatrix} -1 & 0 & -2i \\ 0 & 2 & 0 \\ 2i & 0 & -1 \end{bmatrix}$$

(a) Show that the spectrum of $\hat{P}$ is $\{1, 2, -3\}$.

(b) If the system is in an initial state $|\psi\rangle$, which is represented as $(1/\sqrt{2})(1, 0, 1)^T$, find the probability

i. that a measurement of $\hat{P}$ yields the state $|P = 2\rangle$.

ii. of the event $|\psi\rangle \to |\phi\rangle$ where $|\phi\rangle = (1/\sqrt{2})(|P = 1\rangle + |P = -3\rangle)$.

Chapter 5

Imprecise Measurements and Degeneracy

Back in the second chapter, we introduced the concept of a quantum state as a preparation that results from a measurement. In fact, we claimed that *every* measurement prepares a quantum state. Let us now admit that this is not true in general. In order to give a quick introduction to the basic framework of QM, we had set up a simplified setting. Now I wish to make amends and consider more general scenarios. This will require us to make a slight modification to the mathematical description that we have presented. In particular, we will have to lift the restriction that only linear Hermitian operators having *nondegenerate* eigenvalues can describe observables in QM. Here, we shall see that *every* Hermitian operator will describe some quantum observable and *degeneracy* will play a crucial role in the description.

Precise and Imprecise Measurements

In this section, we shall look at a new type of measurement which we have not discussed so far. The laws of QM as they stand now, will not apply to such measurements. So we will also extend the laws appropriately to describe this new kind of measurement.

Experiments with inadequate resolution

Let us consider an observable $\hat{\alpha}$ having a spectrum $\{\alpha_i \,;\, i = 1, 2, \ldots n\}$. Let this spectrum be the union of the disjoint subsets

$$
\begin{aligned}
A_1 &= \{\alpha_1, \alpha_2, \ldots, \alpha_{g_1}\}, \\
A_2 &= \{\alpha_{g_1+1}, \alpha_{g_1+2}, \ldots, \alpha_{g_1+g_2}\}, \\
\ldots &= \ldots \\
\ldots &= \ldots \\
A_m &= \{\alpha_{g_1+g_2+\ldots+g_{m-1}+1}, \alpha_{g_1+g_2+\ldots+g_{m-1}+2}, \ldots, \alpha_{g_1+g_2+\ldots+g_{m-1}+g_m}\}
\end{aligned}
$$

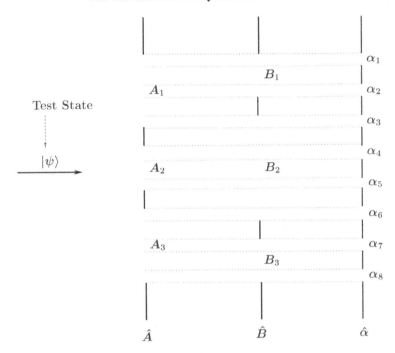

FIGURE 5.1: Imprecise and precise measurements.

(where $\alpha_{g_1+g_2+...+g_{m-1}+g_m} = \alpha_n$). Now imagine an apparatus that is capable of making a measurement that can distinguish between the different disjoint subsets A_i but cannot tell precisely which particular outcome of $\hat{\alpha}$ has actually occurred (see Figure 5.1)[1]. In other words, we are saying that this measurement apparatus has an inadequate resolution for measuring $\hat{\alpha}$. If we like, we could say that the outcomes of the new measurement are the subsets A_i and choose to label the subsets by real numbers (i.e., we can take the $A_i's$ to be real numbers instead of alphabets). We can now associate with our new measurement, a new observable, say $\hat{A}$, with a real spectrum $\{A_i \; ; \; i = 1, 2, \ldots m\}$. It is then only too natural to call the measurement of such an observable $\hat{A}$

---

[1] As a *mnemonic*, imagine the imprecise measurements to be like *filters*. Consider the test state to be a particle moving from left to right whose position along the direction of the slits (call it the y-direction) is not known. Assume that the measurements $\hat{A}$, $\hat{B}$ and $\hat{\alpha}$ are designed to determine the positions along y. Then, if a measurement of $\hat{A}$ leads to an outcome A_1, the measurement of $\hat{\alpha}$ is guaranteed yield a value from the set $\{\alpha_1, \alpha_2, \alpha_3\}$ but A_1 does not specify any definite value of α. Similarly, a measurement of $\hat{B}$ yielding B_3 will only ensure that a subsequent $\hat{\alpha}$ measurement will yield one of the values in $\{\alpha_7, \alpha_8\}$, etc. Note that our picture also suggests that if we perform instantaneously subsequent measurements of the imprecise observables $\hat{A}$ and $\hat{B}$, an outcome A_1 for $\hat{A}$ will guarantee that the outcome of the $\hat{B}$ measurement will yield an outcome from the set $\{B_1, B_2\}$. Although there is nothing to imply this, in what we have said so far in terms of the rules of QM, we shall shortly see that this will indeed turn out to be true.

an *imprecise measurement* of $\hat{\alpha}$. There is, however, one big problem! We have learnt that in QM, *measurements do not really measure* some preexisting value of an observable - they *define* an observable. In this light, it is unclear what meaning one can ascribe to the statement: "a measurement apparatus has inadequate resolution" or that "a measurement is imprecise". So it is important to clarify, precisely, what we intend to mean by these phrases.

Imprecise Measurement *An imprecise measurement corresponding to some precise observable is a measurement process which ensures that*

1. *an instantaneously subsequent measurement of the corresponding precise observable yields a value that belongs to a well defined subset of the spectrum of the precise observable,*

2. *all such subsets associated with different outcomes of the imprecise measurement are disjoint, and*

3. *the union of all such subsets is the spectrum of the precise observable.*

In our example, this means that if a measurement of $\hat{A}$ yields an outcome A_i, a subsequent measurement of $\hat{\alpha}$ is guaranteed to yield a value from the subset $A_i = \left\{ \alpha_{g_1+g_2+...+g_{i-1}+1}, \alpha_{g_1+g_2+...+g_{i-1}+2}, \cdots, \alpha_{g_1+g_2+...+g_{i-1}+g_i} \right\}$. An *imprecise observable* is then defined in terms of imprecise measurements as before.

Imprecise Observable *An imprecise observable is the universal collection of similar imprecise measurements*[2].

If we define imprecise measurement by the criteria outlined above, it naturally brings up the question: how do we know that a measurement is precise? Is it not possible that a precise measurement of an observable is an imprecise measurement with regard to some other observable? For example, how can one be sure that the precise observable $\hat{\alpha}$ in the above example is really precise, and that there does not exist some other hitherto undiscovered observable, say $\hat{\beta}$, with respect to which it is imprecise? We must therefore furnish a criterion for a measurement to be called precise. We move to propose the following definition of a *precise measurement*:

Precise Measurement *A measurement would be called a precise measurement if it prepares a quantum state that is completely specified by the outcome of the measurement.*

This means that after a precise measurement is made on an arbitrary and possibly unknown initial state, subsequent measurements of other observables

---

[2]Recall, that a precise observable was associated with a collection of precise measurements to allow for the fact that an observable can be measured (in the quantum sense) in several ways. The same applies to imprecise measurements.

will yield outcomes with definite probabilities that are completely specified by the outcome of the precise measurement.

The introduction of imprecise measurements forces us to review the laws of QM. We would now have to inquire what the laws of QM could say about imprecise measurements. In particular, we would want to know

1. what is the new state of a quantum system soon after an imprecise measurement has been performed on it (i.e., how the fourth postulate of QM applies to imprecise observables), and

2. whether definite probabilities can be assigned to the different outcomes of an imprecise measurement when the measurement is performed on some well defined initial state (i.e., how the fifth postulate of QM applies to imprecise observables).

We will take up these questions after the next subsection.

Operator describing an imprecise observable

We consider, once again, the precise observable $\hat{\alpha}$ and the corresponding imprecise observable $\hat{A}$. Let us introduce the following notation. We denote the elements of the subset A_i by A_i^r so that the state corresponding to the measurement outcome A_i^r is denoted by $|A_i^r\rangle$. Here i runs from 1 to m and we denote the number of elements in the subset A_i by g_i so that r runs from 1 to g_i for a given i.

Owing to the reality of the labels A_i, and the orthonormality and completeness of the vectors $|A_i^r\rangle$, we can associate a linear Hermitian operator with the imprecise observable $\hat{A}$ using the rule

$$\hat{A}\,|A_i^r\rangle \;=\; A_i\,|A_i^r\rangle$$

in the same way that we associated an operator with a quantum observable earlier. Only this time, by construction, the outcomes A_i's become the *degenerate* eigenvalues of the operator[3] $\hat{A}$. The states $|A_i^r\rangle$ are the eigenvectors of the eigenvalue A_i. Thus, we see that

- *with every imprecise observable one can associate a linear Hermitian operator (acting on the state space) with degenerate eigenvalues.*

As we have done before, we assert that the converse to this statement is also true:

- *with every linear Hermitian operator (acting on the state space) with degenerate eigenvalues one can associate an imprecise observable.*

---

[3]Recall that in the previous instance, when we associated a linear operator with a precise observable, the eigenvalues were *nondegenerate*.

One can ask, given an arbitrary Hermitian operator with degenerate eigenvalues (which would describe some imprecise observable by the above assertion), how does one identify a precise observable associated with it? Well, to do this, one merely has to define an operator by writing down for it, a real spectrum (arbitrarily) whose cardinality is equal to the dimension of the underlying vector space, and then assign to each member of this spectrum (arbitrarily), a unique eigenvector from the orthonormal basis of eigenvectors of the imprecise observable under question.

State resulting from a general measurement

Let us consider, yet again, the precise observable $\hat{\alpha}$ and the corresponding imprecise observable $\hat{A}$ that was used in the previous section. We can expand an arbitrary state $|\psi\rangle$ as

$$
\begin{aligned}
|\psi\rangle &= \sum_{i=1}^{n} |\alpha_i\rangle \langle \alpha_i|\psi\rangle \\
&= \sum_{i=1}^{m} \sum_{r=1}^{g_i} |A_i^r\rangle \langle A_i^r|\psi\rangle
\end{aligned}
$$

where n and m are the total number of elements in the spectra of $\hat{\alpha}$ and $\hat{A}$ respectively, and g_i, as before, represents the number of elements in the subset of the α spectrum that corresponds to A_i. Now, if we a make a measurement of $\hat{\alpha}$ resulting in some outcome A_i^r, it would eliminate the possibility of getting anything other than the outcome A_i^r if the measurement of $\hat{\alpha}$ were repeated instantaneously. We know that the final state would simply become $|A_i^r\rangle$. On the other hand, a measurement of $\hat{A}$ resulting in some outcome A_i would only eliminate the possibility of getting anything other than the elements of the subset labelled by A_i in an instantaneously subsequent measurement of $\hat{\alpha}$. Whereas a measurement of $\hat{\alpha}$ filters out a single outcome, a measurement of $\hat{A}$ will filter out a *set* of outcomes. Thus, *we could regard an imprecise measurement as a coarse filtering process.*

Now it turns out that in addition to coarse filtering, an imprecise measurement has the following property:

- *The relative probabilities of getting different outcomes upon measurement of a precise observable remain unaltered by a preceding measurement of a related imprecise observable, provided that the outcomes of the measurement of the precise observable belong to the same subset associated with the outcome of the imprecise observable.*

That is, if we *tentatively* denote by $|\psi\,;A_i\rangle$, the state resulting from a measurement of the imprecise observable $\hat{A}$ on the state $|\psi\rangle$ that yields an

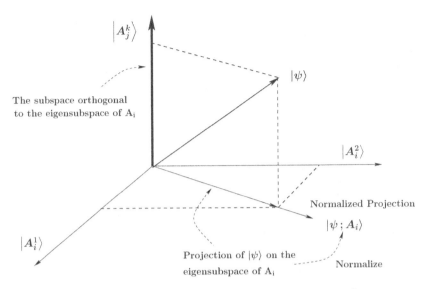

FIGURE 5.2: State Collapse: After an imprecise measurement of $\hat{A}$ on a state $|\psi\rangle$ leading to a doubly degenerate eigenvalue A_i, the final state $|\psi\,;A_i\rangle$ is the normalized projection on the eigenspace of A_i. Here we have again used a real vector space for visualization.

outcome A_i, then we have[4]

$$\frac{P\left[|\psi\,;A_i\rangle \to A_i^r\right]}{P\left[|\psi\,;A_i\rangle \to A_i^s\right]} \;=\; \frac{P\left[|\psi\rangle \to A_i^r\right]}{P\left[|\psi\rangle \to A_i^s\right]}$$

The simplest way to implement this feature in our mathematical framework is to demand that

after making an imprecise measurement the initial state would collapse to the eigenspace associated with the outcome of the measurement.

This means that if the initial state is $\sum_{i=1}^{m}\sum_{r=1}^{g_i}|A_i^r\rangle\langle A_i^r|\psi\rangle$, and a measurement of $\hat{A}$ yields A_i, then the final state will become $\sum_{r=1}^{g_i}|A_i^r\rangle\langle A_i^r|\psi\rangle$. However, such a vector is not normalized and so we must rescale it in order that it may describe a quantum state. This leads to a proposition for the postulate that we are looking for (see Figure 5.2).

---

[4]We say *"tentatively"* because we are actually *anticipating* that a measurement of an imprecise observable results in a *quantum state*. This is also suggested by the notation $|\psi\,;A_i\rangle$. While this is indeed true (as will be stated shortly), the anticipation is not essential. One can simply take the notation $|\psi\,;A_i\rangle \to \beta_j$ to refer to a prospective event that a measurement of an imprecise observable $\hat{A}$ made on a state $|\psi\rangle$, that yields an outcome A_i, upon an instantaneously subsequent measurement of an observable $\hat{\beta}$ yields β_j.

State Collapse Postulate *If on the state $|\psi\rangle$ we make a measurement of an imprecise observable $\hat{A}$ that yields a g_i-fold degenerate eigenvalue A_i as an outcome, we get a new state $|\psi \, ; A_i\rangle$ given by*

$$|\psi \, ; A_i\rangle \;\; = \;\; \frac{\sum_{r=1}^{g_i} |A_i^r\rangle \langle A_i^r|\psi\rangle}{\sqrt{\sum_{r=1}^{g_i} |\langle A_i^r|\psi\rangle|^2}}$$

where the vectors $|A_i^r\rangle$ are a set of orthonormal eigenvectors of $\hat{A}$ corresponding to the eigenvalue A_i.

Of course, finally, whether a law is correct or not is decided by experiment, and it turns out that our proposition is indeed a correct law of nature. It is not hard to see that this law is not the only way to implement the requirement of unaltered relative probabilities. There are other possibilities of a final state which would reproduce this feature. However, every such choice will have distinguishable observable consequences, and therefore will admit the possibility of being ruled out by experiment[5]. This is simply an instance of a situation, quite commonly encountered in physics, where a requirement of mathematical simplicity leads to the correct law of nature.

It may be remarked here that the above rule, that determines what the final state becomes soon after an imprecise measurement, could be used as the definition of an imprecise measurement. However, this would be a bit of an overkill since our earlier definition is less restrictive than the above assertion (as you have been invited to explore in the previous footnote).

Note that the law embodied in the above formula automatically includes the case of precise measurements. If $\hat{A}$ were a precise measurement, all its eigenvalues would be nondegenerate (i.e., $g_i = 1$ for all i). Hence, the final state $|\psi \, ; A_i\rangle$ would simply become $|A_i\rangle$, which is what we expect. In fact, since the eigenspaces are 1-dimensional for precise observables, the collapse to the eigenspace would be most severe for precise measurements.

It might appear that the content of this subsection is in stark contradiction to our earlier statement: "imprecise measurements do not prepare a state". We have just written down the formula that tells us what state is prepared by an imprecise measurement! To understand why this is not a contradiction, one has to note that

an imprecise measurement leads to a well defined state only if the initial state on which the measurement is performed, is known.

An imprecise measurement made on a system whose state is *a priori* unknown, does not prepare a state. Thus, if we start with an imprecise measurement and record its outcome, we will not be able to predict the probability of outcomes

---

[5]Think about the other possibilities.

of subsequent measurements. This is what we mean when we say imprecise measurements do not prepare a state.

When we make a precise measurement, the initial state on which the measurement is performed becomes irrelevant in the specification of the final state, whereas when we make an imprecise measurement, the final state depends crucially on the initial state on which the measurement is performed.

Probability of outcomes of a general measurement

Fortunately, the postulate for computing probability does not need to be altered or extended in order to incorporate imprecise measurements. With the state preparation postulate laid out in the previous subsection, the existing postulate for computing probabilities automatically includes the case of imprecise measurements. To see how, recall that the probability $P\left[|\psi_i\rangle \to |\psi_f\rangle\right]$ of an event characterized by the initial state $|\psi_i\rangle$ and final state $|\psi_f\rangle$ is given by

$$P\left[|\psi_i\rangle \to |\psi_f\rangle\right] = |\langle\psi_f|\psi_i\rangle|^2$$

where $\langle\psi_f|\psi_i\rangle$ is the inner product of the states $|\psi_i\rangle$ and $|\psi_f\rangle$. Now if a measurement of the imprecise observable $\hat{A}$ is made on some state $|\psi\rangle$, the probability for getting some outcome A_i would be given by $P\left[|\psi\rangle \to |\psi; A_i\rangle\right]$. If g_i is the degree of degeneracy of the eigenvalue A_i, and $|A_i^r\rangle$ are the associated g_i eigenvectors (assumed to be orthonormal), we have

$$P\left[|\psi\rangle \to |\psi; A_i\rangle\right] = |\langle\psi; A_i|\psi\rangle|^2$$

$$= \left|\frac{\sum_{r=1}^{g_i}|\langle A_i^r|\psi\rangle|^2}{\sqrt{\sum_{r=1}^{g_i}|\langle A_i^r|\psi\rangle|^2}}\right|^2$$

$$= \sum_{r=1}^{g_i}|\langle A_i^r|\psi\rangle|^2$$

It is evident that if $\hat{A}$ is a precise observable ($g_i = 1$ for all i), the above formula reduces to the familiar result for precise measurements:

$$P\left[|\psi\rangle \to |A_i\rangle\right] = |\langle A_i|\psi\rangle|^2$$

The general rule to compute the probability of a measurement outcome (irrespective of whether the measurement is precise or imprecise) is given by

$$P\left[|\psi\rangle \to A_i\right] = \sum_{k=1}^{g_i}|\langle A_i^k|\psi\rangle|^2$$

where g_i is the degree of degeneracy of the eigenvalue A_i and $\left|A_i^k\right\rangle$ are the associated orthonormal eigenvectors[6].

Postulates of Quantum Mechanics - Version 3

In the preceding sections we have seen how Hermitian operators with degenerate eigenvalues describe imprecise observables, and how the postulate of state preparation has been generalized to accommodate imprecise observables. These developments, clearly, call for another listing of the postulates.

1. Every system is associated with an inner product space. States of the system are the normalized vectors belonging to this space.

2. Every observable of the system is associated with a linear Hermitian operator acting on the state space.

3. The eigenvalues of the Hermitian operator associated with an observable constitute the measurement spectrum of the observable.

4. If a measurement of an observable $\hat{\phi}$ is made on some state $|\psi\rangle$ that yields an eigenvalue ϕ_j, soon after the measurement, the new state becomes $|\psi\,;\,\phi_j\rangle$ which is given by the *collapse postulate*:

$$|\psi\,;\,\phi_j\rangle \;=\; \frac{\sum_{r=1}^{g_j}\left|\phi_j^r\right\rangle\left\langle\phi_j^r|\psi\right\rangle}{\sqrt{\sum_{r=1}^{g_j}\left|\left\langle\phi_j^r|\psi\right\rangle\right|^2}}$$

where $\left|\phi_j^r\right\rangle$ are orthonormal eigenvectors of the eigenvalue ϕ_j and g_j is its degree of degeneracy.

5. The probability $P\left[|\psi\rangle \to |\phi\rangle\right]$ of an event $|\psi\rangle \to |\phi\rangle$ is given by the *Born rule*:

$$P\left[|\psi\rangle \to |\phi\rangle\right] \;=\; \left|\langle\phi|\psi\rangle\right|^2$$

where the scalar product $\langle\phi|\psi\rangle$ is called the probability amplitude of the event $|\psi\rangle \to |\phi\rangle$.

---

[6]Note that despite the resemblance, this is not the classical probability addition rule. The event $\psi \to A_i$ is not defined to be the collection of elementary events $\left\{|\psi\rangle \to \left|A_i^r\right\rangle\right\}$. The values A_i are actual outcomes of an actual measurement process. Incidentally, although the meaning of the notation $P\left[|\psi\rangle \to A_i\right]$ is self explanatory, we must admit that we have not defined it explicitly. For the puritan, the $|\psi\rangle \to A_i$ that P refers to may be taken to be the (quantum) event $|\psi\rangle \to |\psi; A_i\rangle$, which is characterized by the initial state $|\psi\rangle$ and final state $|\psi\,;\,A_i\rangle$.

Problems

1. A state $|\psi\rangle$ is expanded in the orthonormal basis of eigenvectors $\{|q_1\rangle, |q_2\rangle, |q_3\rangle\}$ of a certain observable $\hat{Q}$ as

$$|\psi\rangle = \frac{1}{\sqrt{2}}|q_1\rangle + \frac{1}{2}|q_2\rangle + \frac{1}{2}|q_3\rangle$$

The states $|q_1\rangle$ and $|q_2\rangle$ correspond to a 2-fold degenerate eigenvalue η of $\hat{Q}$. If the observable $\hat{Q}$ is measured on $|\psi\rangle$ what is probability that the outcome will be η and what will be the final state in this case?

2. Let $\hat{G}$ be an observable having a degenerate eigenvalue μ. The eigenspace of μ is spanned by the vectors $(0, 1, 1)^T$ and $(0, 2, -1)^T$. If $\hat{G}$ is measured on the state $(1/\sqrt{2})(1, 1, 0)^T$, what will be the state soon after the measurement if the outcome is μ, and what is the probability for this event?

3. For a certain three level system, two observables $\hat{P}$ and $\hat{Q}$ are represented by

$$\left[\hat{P}\right] = p\begin{bmatrix} 1 & 0 & 0 \\ 0 & 0 & 1 \\ 0 & 1 & 0 \end{bmatrix} \quad \text{and} \quad \left[\hat{Q}\right] = q\begin{bmatrix} 4 & 0 & 0 \\ 0 & 7 & -1 \\ 0 & -1 & 7 \end{bmatrix}$$

where p and q are dimensionful constants.

(a) Show that $\hat{P}$ is an imprecise observable in regard to $\hat{Q}$.

(b) If a measurement of $\hat{P}$ is made on states $|\psi\rangle$ and $|\phi\rangle$ represented respectively by

$$[\psi] = \frac{1}{\sqrt{3}}\begin{bmatrix} 1 \\ 1 \\ 1 \end{bmatrix} \quad \text{and} \quad [\phi] = \frac{1}{\sqrt{14}}\begin{bmatrix} 1 \\ 2 \\ 3 \end{bmatrix}$$

what is the probability of getting an outcome p?

(c) Write down the final state $|\psi\,;\,p\rangle$.

(d) If a measurement $\hat{Q}$ is made on $|\psi\,;\,p\rangle$ find out the probabilities of the different possibilities.

(e) As you may have found out, the spectrum of $\hat{Q}$ is $\{2q, 3q, 4q\}$. The imprecise observable $\hat{P}$ resolves between the subsets $\{2q, 3q\}$ and $\{4q\}$ of the $\hat{Q}$-spectrum. Construct another imprecise observable that will resolve between the following partitioning of the $\hat{Q}$-spectrum: $\{2q\}$ and $\{3q, 4q\}$. Express this observable in the same basis in which $\hat{P}$ and $\hat{Q}$ have been represented.

4. Check that after a measurement of an imprecise observable, an instantaneously subsequent measurement of the corresponding precise observable can only lead to an outcome that belongs to the subset associated with the outcome of the imprecise measurement.

5. Show that if a measurement of an imprecise observable is instantaneously repeated, the outcome will be reproduced:

$$P\left[|\psi\,;A_i\rangle \to A_j\right] = \delta_{ij}$$

6. If $\hat{P}_i$ is the projection operator associated with the eigenvalue A_i of the observable $\hat{A}$, show that

   (a) the postulate of QM that specifies the final state soon after a measurement of $\hat{A}$ is made on a state $|\psi\rangle$ that yields the outcome A_i (*collapse postulate*) can be written as

   $$|\psi\,;A_i\rangle = \frac{\hat{P}_i\,|\psi\rangle}{\sqrt{\left\langle\psi|\hat{P}_i|\psi\right\rangle}}$$

   (b) the postulate of QM that furnishes the probability of getting the outcome A_i upon a measurement of $\hat{A}$ on a state $|\psi\rangle$ (*Born rule*) can be written as

   $$P\left[|\psi\rangle \to A_i\right] = \left\langle\psi|\hat{P}_i|\psi\right\rangle$$

7. A certain system is described by the inner product space V. A precise observable $\hat{a}$ acts on V and has a spectrum $\{\alpha_i\,;\,i = 1, 2, \ldots, 8\}$. Now consider three observables $\hat{A}$, $\hat{B}$ and $\hat{C}$ which are imprecise with respect to $\hat{a}$. The respective spectra $\{A_i\}$, $\{B_j\}$ and $\{C_k\}$ of the observables $\hat{A}$, $\hat{B}$ and $\hat{C}$ and associated subsets of the spectrum of $\hat{a}$ are given by

$$
\begin{aligned}
A_1 &= \{\alpha_1, \alpha_2, \alpha_3, \alpha_4\} \\
A_2 &= \{\alpha_5, \alpha_6, \alpha_7, \alpha_8\} \\
B_1 &= \{\alpha_1, \alpha_2, \alpha_5, \alpha_6\} \\
B_2 &= \{\alpha_3, \alpha_4, \alpha_7, \alpha_8\} \\
C_1 &= \{\alpha_1, \alpha_3, \alpha_5, \alpha_7\} \\
C_2 &= \{\alpha_2, \alpha_4, \alpha_6, \alpha_8\}
\end{aligned}
$$

   (a) Show that the final state resulting from instantaneously subsequent measurements of $\hat{A}$ and $\hat{B}$ is independent of the order of the measurement:

   $$|\psi;\,A_i,\,B_j\rangle = |\psi;\,B_j,\,A_i\rangle$$

(b) Using the previous result, show that

$$P\left[|\psi; A_i, B_j\rangle \to A_k\right] = \delta_{ik}$$

(c) Show that instantaneously subsequent measurements of the imprecise observables $\hat{A}$, $\hat{B}$ and $\hat{C}$ results in a final state that is independent of the initial state:

$$|\psi; A_i, B_j, C_k\rangle = |\phi; A_i, B_j, C_k\rangle = |\chi; A_i, B_j, C_k\rangle = \dots$$

where $|\psi\rangle, |\phi\rangle, |\chi\rangle, \dots$, etc., are arbitrary initial states. The final states can then simply be written as $|A_i, B_j, C_k\rangle$.

(d) It is not essential, in general, that the partitioning of the spectrum of a precise observable by a set of imprecise observables will be such that it admits all possible combinations of outcomes of the imprecise observables (like the case just considered). Can you imagine a partitioning of the α spectrum provided by some pair of imprecise observables such that all combinations of outcomes are naturally forbidden?

Complete and Incomplete Measurements

We know that if we make two instantaneously subsequent *precise* measurements, the second measurement wipes out the memory of the first. If the first measurement *prepares* a system in some state that we label by the outcome of the measurement, then the second measurement *destroys* this state and prepares a *new* state that is now completely specified by the outcome of the second measurement. This, however, is not true if *imprecise* measurements are involved. For example, imagine that we follow up a *precise* measurement with a second measurement that is *imprecise*, but with regard to some *different* precise observable. Then the new state that will be prepared by the imprecise measurement, will carry an imprint of the earlier state prepared by the precise measurement. This new state cannot be completely specified by the outcome of the imprecise measurement alone; one must also use the outcome of the previous precise measurement. A series of imprecise measurements, made on some well defined initial state, will in general, produce a state that will carry residual footprints of all the previous measurements in the history of its preparation starting from the precise measurement that prepared the initial state. Memories of previous measurements are not, in general, wiped out by imprecise measurements. Thus, it appears that in order to prepare a quantum state that is independent of the history of past measurements, one must necessarily make a precise measurement on the system.

But, if you have worked out the last problem carefully, you will have seen a counter example: a set of measurements, that were imprecise in regard to the *same* precise observable, could actually *collaborate* to prepare a quantum state specified completely in terms of the set of outcomes of the sequence of imprecise measurements. A suitable sequence of instantaneously subsequent imprecise measurements can indeed wipe out the memory of the initial state. To discuss this in a general context, it is useful to introduce the following definitions.

Compatible Observables   *A set of observables are said to be compatible if there exists an orthonormal basis consisting of common eigenvectors of the observables*[7].

It is clear that

1. *a precise observable is compatible with all its imprecise observables,* and

2. *observables that are imprecise with respect to the same precise observable are mutually compatible.*

The following important properties of compatible observables are easy to prove:

- *The final state resulting from instantaneously subsequent measurements of a set of compatible measurements is independent of the order of the measurements.*

- *After a set of instantaneously successive measurements of compatible observables, if a measurement belonging to the set is repeated instantaneously, its previous outcome will be reproduced with certainty*[8].

Thus, if $\hat{A}$ and $\hat{B}$ are compatible observables, then for all measurement outcomes A_i, B_j of $\hat{A}$ and $\hat{B}$ respectively, we have

$$|\psi ; A_i, B_j\rangle = |\psi ; B_j, A_i\rangle ,$$
$$P\left[|\psi ; A_i, B_j\rangle \to A_i\right] = 1 = P\left[|\psi ; B_j, A_i\rangle \to B_j\right]$$

for an arbitrary initial state $|\psi\rangle$.

Incidentally, if a set of observables are mutually compatible (i.e., compatible in pairs), then there exists an orthonormal basis of their common eigenvectors, and the set is said to be compatible as a whole. The simplest way to prove this uses results discussed in the next section[9].

---

[7]Actually it is adequate to require that there exists a *complete set* of common eigenvectors of the observables, but it turns out that in this case, an *orthonormal* basis of common eigenvectors can always be found.

[8]It is, obviously, this characteristic that makes us call compatible observables "compatible"; they *tolerate* each other instead of wiping each other out.

[9]One might wonder if by using these properties, it is possible to decide experimentally

Simultaneous Measurement *Instantaneously subsequent measurements of compatible observables are called simultaneous measurements.*

Note that we do not really mean that the measurements are simultaneous in *time*. We call them simultaneous because of the *simultaneous certainty* of reproducing the outcomes upon instantaneously subsequent measurements. In some sense, the *information about the outcomes exist simultaneously* in the final state.

Complete Set of Compatible Observables *If simultaneous measurement of all members of a set of compatible observables always prepares a state that is independent of the initial state, then the set of compatible observables is said to form a complete set of compatible observables (CSCO).*

In other words, a set of observables forms a CSCO if the *memory of the initial state is completely wiped out* after simultaneous measurements of *all* the observables of the set (irrespective of the outcomes of the simultaneous measurements). Thus, if $|\psi\rangle, |\phi\rangle, |\chi\rangle, \ldots$ are arbitrary initial states and the set of observables $\left\{\hat{A}, \hat{B}, \hat{C}, \ldots\right\}$ comprises a CSCO, then

$$|\psi\,;\,A_i, B_j, C_k, \ldots\rangle = |\phi\,;\,A_i, B_j, C_k, \ldots\rangle = |\chi\,;\,A_i, B_j, C_k, \ldots\rangle = \ldots$$

where $A_i, B_j, C_k, \ldots$ are arbitrary outcomes of simultaneous measurements of $\hat{A}, \hat{B}, \hat{C}, \ldots$ respectively. Since the *initial state becomes irrelevant* in the specification of the final state, it makes no sense to carry the burden of the labels $\psi, \phi, \chi, \ldots$ that indicate the initial state. We can denote the final state simply as $|A_i, B_j, C_k, \ldots\rangle$. A set of compatible observables will therefore form a CSCO if after a simultaneous measurement of *all* the observables of the set,

---

whether a set of observables are compatible. Indeed, it turns out that either of these properties is *completely equivalent* to the definition of compatibility. However, one has to be cautious! In order to decide the compatibility, it may be tempting, for example, to test for two *prospective* compatible observables $\hat{A}$ and $\hat{B}$, whether the measurement outcomes are appropriately sustained:

$$P\left[|\psi; A_i, B_j\rangle \to A_i\right] = 1 = P\left[|\psi; B_j, A_i\rangle \to B_j\right]$$

for all outcomes A_i and B_j of $\hat{A}$ and $\hat{B}$ respectively. But, one should bear in mind that this test, performed on a particular state $|\psi\rangle$, or even a full basis of states, cannot ascertain compatibility. If the sustainability of measurement outcomes, referred to above, is *not* maintained, we can, at once, say that the observables are incompatible. However, if the measurement outcomes *are* sustained, it will, unfortunately, not be enough to claim that the observables are compatible. It will merely be an indication that they might be. If one performs the test on *all possible states* of the system (which is clearly impossible in practice), and gets an affirmative result, only then can compatibility be inferred. Similarly, it is also possible to check compatibility by testing (which is, again, impossible in practice), if the state resulting from instantaneously subsequent measurement of the observables is independent of the order of the measurements for all possible initial states, and all possible measurement outcomes.

the final state is, always, completely specified by the set of their measurement outcomes[10]. Such a set of measurements is often called a *complete set of measurements* in QM.

It is clear that

a precise observable will constitute a CSCO by itself but imprecise observables must team up in order to form a CSCO.

The preceding concepts lead to the following terminology:

Complete and Incomplete Measurements *A simultaneous measurement of a complete set of compatible observables is often referred to as a complete measurement. Similarly, simultaneous measurement of a set of compatible observables which do not form a CSCO is called an incomplete measurement.*

If a CSCO comprises N observables $\left\{ \hat{A}, \hat{B}, \hat{C}, ... \right\}$, then the states prepared by the simultaneous measurement of this CSCO will be described by N parameters: $\{A_i, B_j, C_k, ...\}$ where A_i, B_j, C_k, ... are elements of the spectra of the observables $\hat{A}$, $\hat{B}$, $\hat{C}$, ... respectively[11]. We shall often, for the sake of brevity, use a single label to stand for the complete set of measurement outcomes (corresponding to a CSCO) that specifies a state. Thus, we shall often write a state such as $|A_i, B_j, C_k, ...\rangle$ as $|\psi\rangle$ where ψ will collectively denote the labels $A_i, B_j, C_k, ...$.

Until the last chapter, we have been working under the *unnatural assumption* that the spectra of all observables have the same number of elements, which is equal to the dimension of the state space. It is obvious that this assumption will now apply only to precise observables. With imprecise observables introduced in the picture, the situation becomes much more natural.

The spectra of observables are allowed to have any size. If the number of elements in the spectrum of an observable is equal to the dimension of the

---

[10]To find a CSCO, we can imagine that we go on making instantaneously subsequent measurements of compatible observables until we reach a point when probabilities for all possible outcomes of measurement of an arbitrary (and possibly incompatible) observable becomes independent of the initial state. When this happens we know that we have prepared a well defined state, and this state can then be labelled by the outcomes of the compatible measurements. If this is true for all possible outcomes of the set of compatible measurements, we know that the set of compatible measurements corresponds to a CSCO. This procedure is, obviously, impossible to carry out in practice but it underlies the meaning of a CSCO. A practical method will be discussed at the end of this chapter.

[11]One often loosely refers to the number of elements in the CSCO as the *dimension of the system*. This is not to be confused with the dimension of the inner product space that constitutes the state space of the system. We say "loosely" because this number is really not a characteristic of the system but merely a characteristic of the description: the choice of the CSCO. For a given system it is not necessary that every CSCO will have the same number of elements. Actually, it is trivially possible to treat even a 1-dimensional system as higher dimensional and *vice versa*.

state space, the observable is precise, else it is imprecise.

It is not difficult to see that if a set of observables $\left\{ \hat{A}, \hat{B}, \hat{C}, \ldots \right\}$ forms a CSCO, then the set of states $|A_i, B_j, C_k, \ldots\rangle$ with $(A_i, B_j, C_k, \ldots)$ running over all *permissible* combinations of the measurement outcomes of $\hat{A}, \hat{B}, \hat{C}, \ldots$ forms an orthonormal basis in the state space[12]. Since they will often be used in QM, let us acquaint ourselves with the most important formulae involving such bases.

In the following equations, for all the summations, the indices have been assumed to run over all their allowed values.

Orthonormality of the basis states read

$$\langle A_{i'}, B_{j'}, C_{k'}, \ldots | A_i, B_j, C_k, \ldots \rangle \quad = \quad \delta_{ii'} \delta_{jj'} \delta_{kk'} \ldots$$

An arbitrary vector $|\psi\rangle$ can be written as

$$|\psi\rangle \quad = \quad \sum_{i,j,k,\ldots} |A_i, B_j, C_k, \ldots\rangle \langle A_i, B_j, C_k, \ldots | \psi\rangle$$

If $|\psi\rangle$ is a state vector, then the totality condition obeyed by the amplitudes $\langle A_i, B_j, C_k, \ldots | \psi\rangle$ will take the form

$$\sum_{i,j,k,\ldots} |\langle A_i, B_j, C_k, \ldots | \psi\rangle|^2 = 1$$

The formula for the inner product $\langle \phi | \psi \rangle$ will appear as

$$\langle \phi | \psi \rangle \quad = \quad \sum_{i,j,k,\ldots} \langle \phi | A_i, B_j, C_k, \ldots\rangle \langle A_i, B_j, C_k, \ldots | \psi\rangle$$

If we have an initial state

$$|\psi\rangle \quad = \quad \sum_{i,j,k,\ldots} |A_i, B_j, C_k, \ldots\rangle \langle A_i, B_j, C_k, \ldots | \psi\rangle$$

on which we make a measurement of an observable $\hat{A}$ belonging to the CSCO $\left\{ \hat{A}, \hat{B}, \hat{C}, \ldots \right\}$ that yields an outcome $A_{i'}$, soon after the measurement the new state $|\psi\,;\,A_{i'}\rangle$ will be given by

$$|\psi\,;\,A_{i'}\rangle \quad = \quad \frac{\sum_{j,k,\ldots} |A_{i'}, B_j, C_k, \ldots\rangle \langle A_{i'}, B_j, C_k, \ldots | \psi\rangle}{\sqrt{\sum_{j,k,\ldots} |\langle A_{i'}, B_j, C_k, \ldots | \psi\rangle|^2}}$$

---

[12]It is possible, and quite often the case that every possible combination of the measurement outcomes of a set of compatible observables *does not* correspond to an allowed state.

The probability of getting an outcome $A_{i'}$ in a measurement of the observable $\hat{A}$ made on a state $|\psi\rangle$ will be given by

$$P\left[|\psi\rangle \to |\psi\,;\, A_{i'}\rangle\right] \;=\; \sum_{j,k,\dots} |\langle A_{i'}, B_j, C_k, \dots |\psi\rangle|^2$$

Along the same lines, the probability of getting the outcomes $A_{i'}$ and $B_{j'}$ in a simultaneous measurement of the observables $\hat{A}$ and $\hat{B}$ made on the state $|\psi\rangle$ would be

$$P\left[|\psi\rangle \to |\psi\,;\, A_{i'}, B_{j'}\rangle\right] \;=\; \sum_{k,\dots} |\langle A_{i'}, B_{j'}, C_k, \dots |\psi\rangle|^2$$

By repeated application of the same procedure, the probability of getting the outcomes $A_{i'}, B_{j'}, C_{k'}, \dots$ in a simultaneous measurement of all the members of a CSCO $\left\{\hat{A}, \hat{B}, \hat{C}, \dots\right\}$ made on $|\psi\rangle$ will be

$$P\left[|\psi\rangle \to |\psi\,;\, A_{i'}, B_{j'}, C_{k'}, \dots\rangle\right] \;=\; |\langle A_{i'}, B_{j'}, C_{k'}, \dots |\psi\rangle|^2$$

All these formulae are elementary consequences of the definitions provided in this section.

In real life, to see whether a set of imprecise observables forms a CSCO, one does not usually check whether the defining criteria are satisfied. It is also seldom necessary to identify the underlying precise observable (although such an observable can be trivially constructed *a posteriori*). At the level of quantization, one usually specifies a first basis in terms of the spectra of a set of imprecise observables which are *assumed* to form a CSCO[13]. Testing the compatibility of new observables and construction of new CSCO's are then done using purely algebraic properties of the operators that describe compatible observables. This is what we intend to discuss now.

Commuting Operators

The ***commutator*** $\left[\hat{A}, \hat{B}\right]$ of two operators $\hat{A}$ and $\hat{B}$ is defined by

$$\left[\hat{A}, \hat{B}\right] = \hat{A}\hat{B} - \hat{B}\hat{A}$$

The operators $\hat{A}$ and $\hat{B}$ are said to ***commute*** if they have a vanishing commutator. Now, let us introduce some properties of such operators.

---

[13] This is, of course, guided by some physical input or intuition.

- *If two linear operators have a complete set of common eigenvectors then the operators must commute.*

It is extremely easy to prove this result. Let $\{|\chi_i\rangle \,;\, i = 1, 2, \ldots, n\}$ be a complete set of common eigenvectors of two linear operators $\hat{A}$ and $\hat{B}$ acting on a vector space V. We can write an arbitrary vector $|\psi\rangle$ in V as

$$|\psi\rangle = \sum_{i=1}^{n} \psi^i |\chi_i\rangle$$

where the expansion coefficients ψ^i are appropriate scalars. Now, owing to the linearity of the operators $\hat{A}$ and $\hat{B}$, we have

$$\hat{A}\hat{B}|\psi\rangle = \sum_{i=1}^{n} \psi^i A_i B_i |\chi_i\rangle = \sum_{i=1}^{n} \psi^i B_i A_i |\chi_i\rangle = \hat{B}\hat{A}|\psi\rangle$$

Here A_i and B_i are the eigenvalues of $\hat{A}$ and $\hat{B}$ corresponding to the common eigenvector $|\chi_i\rangle$. Since $|\psi\rangle$ is arbitrary, it follows that

$$\hat{A}\hat{B} = \hat{B}\hat{A}$$

Now let us state the converse to the above theorem.

- *If two diagonalizable linear operators commute, then they will necessarily have a complete set of common eigenvectors*[14].

A straightforward corollary to the above theorem follows.

- *If two Hermitian operators commute, then there must exist an orthonormal basis comprising common eigenvectors of the operators.*

Since in QM, we shall only use the corollary, we shall demonstrate the proof of the corollary directly. This proof will serve as a warm up exercise for an upcoming analysis that we shall shortly undertake.

Let $\hat{A}$ and $\hat{B}$ be two commuting Hermitian operators acting on a vector space V. Let the vectors $\left|A_i^j\right\rangle$ constitute an orthonormal basis of eigenvectors of $\hat{A}$ (since $\hat{A}$ is Hermitian, such a basis will always exist). We assume

$$\hat{A}\left|A_i^j\right\rangle = A_i \left|A_i^j\right\rangle$$

for all j running from 1 to g_i where g_i is the degree of degeneracy of the eigenvalue A_i of $\hat{A}$. Now, owing to the commutativity of $\hat{A}$ and $\hat{B}$

$$\hat{A}\left(\hat{B}\left|A_i^j\right\rangle\right) = \hat{B}\left(\hat{A}\left|A_i^j\right\rangle\right) = \hat{B}\left(A_i\left|A_i^j\right\rangle\right) = A_i\left(\hat{B}\left|A_i^j\right\rangle\right)$$

---

[14]Note that we are not saying that if the two operators have a complete set of eigenvectors (i.e., they are diagonalizable), the eigenvectors will be common (this will happen only if all eigenvalues are nondegenerate, as we shall see). We are only claiming that a complete set of common eigenvectors will *exist*.

Thus $\hat{B}\left|A_i^j\right\rangle$ is also an eigenvector of $\hat{A}$ with the eigenvalue A_i, and it will therefore belong to the eigensubspace associated with A_i so that

$$\hat{B}\left|A_i^j\right\rangle = \sum_{k=1}^{g_i} B_i^{kj}\left|A_i^k\right\rangle$$

This is often expressed by saying that the subspace, say V_i, associated with the eigenvalue A_i is *invariant* under the action of $\hat{B}$. Hence, corresponding to every eigenvalue A_i of $\hat{A}$, we can imagine an operator $\hat{B}_i$ acting on the eigenspace V_i of A_i which may be called the *restriction* of $\hat{B}$ to V_i. The representation of $\hat{B}_i$ in the basis consisting of the vectors $\left|A_i^j\right\rangle$, where i is now fixed and j runs from 1 to g_i, is given by the expansion coefficients

$$B_i^{kj} = \left\langle A_i^k|\hat{B}|A_i^j\right\rangle$$

Owing to the Hermiticity of $\hat{B}$, the restriction $\hat{B}_i$ will also be Hermitian. Hence, there will exist an orthonormal basis of eigenvectors of $\hat{B}_i$ (and therefore, also of $\hat{B}$) in V_i. But all such eigenvectors, being linear combinations of the vectors $\left|A_i^j\right\rangle$ (with i fixed), will continue to be eigenvectors of $\hat{A}$ with the eigenvalue A_i. By repeating this argument in every eigenspace V_i, we can see that an orthonormal basis of common eigenvectors of $\hat{A}$ and $\hat{B}$ will exist.

Complete Set of Commuting Operators *If every common eigenvector of a set of commuting operators can be uniquely specified by specifying the eigenvalues of all the operators of the set, then we say that the set constitutes a complete set of commuting operators.*

Fortunately, a complete set of commuting operators can also be acronymed CSCO (i.e., the same acronym as for a complete set of compatible observables[15]).

In our discussion, a complete set of commuting operators will always refer to a complete set of commuting, *Hermitian* operators.

Construction of a CSCO

To construct a CSCO, consider two commuting Hermitian operators $\hat{A}$ and $\hat{B}$. If the eigenvalues of $\hat{B}_i$ (restriction of $\hat{B}$ to the eigenspace V_i associated with eigenvalue A_i of $\hat{A}$) are *nondegenerate* for *every* V_i, then the set $\left\{\hat{A}, \hat{B}\right\}$ will form a CSCO. Note that we do not need the eigenvalues of the full, *unrestricted* operator $\hat{B}$ to be nondegenerate. It is fine if eigenvectors of $\hat{B}$ belonging to different V_i have the same eigenvalue. We only require the eigenvectors of

---

[15]Can you guess, why this is fortunate?

$\hat{B}$ belonging to the same V_i to have distinct eigenvalues. This is equivalent to requiring that the *restrictions* $\hat{B}_i$ of the operator $\hat{B}$ have nondegenerate eigenvalues. If this is not the case, then for each eigenvalue B_j of $\hat{B}_i$, there will exist an orthonormal set of eigenvectors $|B_{ij}^k\rangle$ which are common eigenvectors of $\hat{A}$ and $\hat{B}$ with eigenvalues A_i and B_j respectively. We can now take another Hermitian operator $\hat{C}$ that commutes with the operators $\hat{A}$ and $\hat{B}$. It is clear that if $\hat{C}$ acts on a common eigenvector $|B_{ij}^k\rangle$ of $\hat{A}$ and $\hat{B}$, with eigenvalues A_i and B_j respectively, then it will give a vector that will belong to the eigenspace associated with A_i as well as B_j:

$$\hat{A}\left(\hat{C}\,|B_{ij}^k\rangle\right) = A_i\left(\hat{C}\,|B_{ij}^k\rangle\right)$$

$$\hat{B}\left(\hat{C}\,|B_{ij}^k\rangle\right) = B_j\left(\hat{C}\,|B_{ij}^k\rangle\right)$$

This means that $\hat{C}\,|B_{ij}^k\rangle$ will belong to the intersection of the eigenspaces associated with A_i and B_j. Let us denote this space by V_{ij} and let g_{ij} be its dimension[16]. Clearly the set of vectors $|B_{ij}^k\rangle$ will form an orthonormal basis in V_{ij}. We shall, as before, be able to define a restriction $\hat{C}_{ij}$ of $\hat{C}$ to the space V_{ij} corresponding to every pair of eigenvalues A_i and B_j of $\hat{A}$ and $\hat{B}$. The action of $\hat{C}_{ij}$ on V_{ij} will be given by

$$\hat{C}_{ij}\,|B_{ij}^k\rangle = \sum_{l=1}^{g_{ij}} C_{ij}^{lk}\,|B_{ij}^l\rangle$$

Like before, owing to the Hermiticity of $\hat{C}$, we shall be able to find an orthonormal basis of eigenvectors of $\hat{C}_{ij}$ (and of $\hat{C}$) in V_{ij}. If the eigenvalues of $\hat{C}_{ij}$ are nondegenerate then the set of operators $\left\{\hat{A},\hat{B},\hat{C}\right\}$ will form a CSCO.

Note, once again, that we do not need the eigenvalues of the operator $\hat{C}$ to be nondegenerate. It is alright if eigenvectors of $\hat{C}$ belonging to different V_{ij} have the same eigenvalue. We only require eigenvectors belonging to the same V_{ij} to have distinct eigenvalues. If some of the eigenvalues of $\hat{C}_{ij}$ are degenerate, we shall need to look for another Hermitian operator, say $\hat{D}$, commuting with every member of $\left\{\hat{A},\hat{B},\hat{C}\right\}$ and see if all eigenvalues of the appropriate restrictions of $\hat{D}$ are nondegenerate. This process can be continued until we find a CSCO.

Illustration

In Table 5.1, we illustrate the construction of a CSCO comprising four observables in an 8-dimensional vector space V. We have denoted the eigenvectors without the '$|\ \rangle$' symbols in the table entries to avoid clutter.

---

[16]Convince yourself that the intersection of two subspaces is also a subspace. Will the union of two subspaces be a vector space?

TABLE 5.1: Construction of a CSCO.

V								
A_1^1	A_1^2	A_1^3	A_1^4	A_2^1	A_2^2	A_3^1	A_3^2	A_3^3
V_1				V_2		V_3		
B_{11}^1	B_{12}^1	B_{12}^2	B_{12}^3	B_{21}^1	B_{23}^1	B_{34}^1	B_{35}^1	B_{35}^2
V_{11}	V_{12}			V_{21}	V_{23}	V_{34}	V_{35}	
C_{111}^1	C_{121}^1	C_{122}^1	C_{122}^2	C_{213}^1	C_{233}^1	C_{343}^1	C_{353}^1	C_{354}^1
V_{111}	V_{121}	V_{122}		V_{213}	V_{233}	V_{343}	V_{353}	V_{354}
D_{1111}^1	D_{1211}^1	D_{1221}^1	D_{1222}^1	D_{2133}^1	D_{2333}^1	D_{3434}^1	D_{3534}^1	D_{3544}^1
V_{1111}	V_{1211}	V_{1221}	V_{1222}	V_{2133}	V_{2333}	V_{3434}	V_{3534}	V_{3544}

1. In a vector space V, the vectors $\left\{ \left| A_i^j \right\rangle \right\}$ comprise a basis of eigenvectors of $\hat{A}$ where $\hat{A} \left| A_i^j \right\rangle = A_i \left| A_i^j \right\rangle$. The eigenspaces associated with the eigenvalues A_i are denoted by V_i. There are three degenerate eigenvalues A_1, A_2 and A_3 with degrees of degeneracy four, two and three respectively.

2. The operator $\hat{B}$ commutes with $\hat{A}$ so that the eigenspaces V_i are invariant under the action of $\hat{B}$, and $\hat{B}$ is completely defined within each eigenspace V_i. We define the operators $\hat{B}_i$ to denote the restrictions of $\hat{B}$ to V_i. The vectors $\left\{ \left| B_{ij}^k \right\rangle \right\}$ comprise a basis of common eigenvectors of $\hat{A}$ and $\hat{B}$, where $\hat{A} \left| B_{ij}^k \right\rangle = A_i \left| B_{ij}^k \right\rangle$ and $\hat{B} \left| B_{ij}^k \right\rangle = B_j \left| B_{ij}^k \right\rangle$. The intersection of the eigenspaces of A_i and B_j are denoted by V_{ij}. We note that only V_{12} and V_{35} have dimension more than one. There are five distinct eigenvalues of $\hat{B}$ denoted by B_1, B_2, B_3, B_4 and B_5 of which B_1, B_2 and B_5 are degenerate. However, only B_2 and B_5 are also degenerate eigenvalues of the restrictions of $\hat{B}$ (to V_1 and V_3). The eigenvalue B_1 is a degenerate eigenvalue of $\hat{B}$ acting on V but not of the restrictions of $\hat{B}$ to any of the eigenspaces V_i. At this point, the vectors belonging to V_{12} and V_{35} cannot be uniquely identified by specifying the eigenvalues of $\hat{A}$ and $\hat{B}$.

3. Now we introduce a third operator $\hat{C}$ which commutes with both, $\hat{A}$ and $\hat{B}$ having a basis of common eigenvectors $\left\{ \left| C_{ijk}^l \right\rangle \right\}$ of $\hat{A}$, $\hat{B}$ and $\hat{C}$. We need to check if all eigenvalues of the restrictions of $\hat{C}$ to V_{12} and V_{35} are nondegenerate. This will ensure that the basis of common eigenvectors $\left\{ \left| C_{ijk}^l \right\rangle \right\}$ of $\hat{A}$, $\hat{B}$ and $\hat{C}$ will be uniquely specified by the eigenvalues

of $\hat{A}$, $\hat{B}$ and $\hat{C}$. We see that this happens for V_{35} but not for V_{12}, which still has a degenerate eigenvalue C_2 having an associated subspace V_{122} (being the intersection of the eigenspaces of A_1, B_2 and C_2) of dimension two.

4. Finally, we introduce the operator $\hat{D}$ that commutes with $\hat{A}$, $\hat{B}$ and $\hat{C}$ and whose restriction to V_{122} does have all nondegenerate eigenvalues, completing the construction of the CSCO $\left\{\hat{A}, \hat{B}, \hat{C}, \hat{D}\right\}$.

Compatibility and Commutativity

The mathematical results of the previous section shows the obvious connection between compatible observables and commutating Hermitian operators:

- *If a set of observables are compatible, then the corresponding operators which describe them must commute with each other.*

and conversely

- *If a set of Hermitian operators commute among themselves, then the observables, that the Hermitian operators correspond to, are compatible.*

Moreover, the definition of a complete set of commuting operators makes the following connection evident:

- *A complete set of compatible observables is described by a complete set of commuting Hermitian operators and vice versa.*

In actual practice, these are the connections that are used to identify compatible observables and their complete sets.

Uncertainty principle

In the early days of QM, people constantly had to be reminded that QM is a probabilistic theory as opposed to the deterministic ones that were known up until the arrival of QM. Arguably, the most famous result that has played the role of this reminder is the so called *uncertainty principle*, due to Werner Heisenberg, one of the founding fathers of QM[17]. It states that the product of the uncertainties of two incompatible observables (essentially arising out of being simultaneously indeterministic) has a lower bound. The exact value of the lower bound depends on the estimate that is chosen to express the uncertainty. If the standard deviation (square root of the expectation value

---

[17]Deservingly, it is commonly referred to as *Heisenberg's uncertainty principle*.

of the square of the deviation from the mean) is taken as a measure of the uncertainty, then the uncertainty principle reads

$$\sigma_A^2 \sigma_B^2 \geq \left(\frac{1}{2i} \left\langle \left[\hat{A}, \hat{B} \right] \right\rangle \right)^2$$

where $\hat{A}$ and $\hat{B}$ are two incompatible observables having a commutator $\left[\hat{A}, \hat{B} \right]$ and

$$\sigma_X = \sqrt{\left\langle (X - \langle X \rangle)^2 \right\rangle}$$

is the standard deviation of the observable[18] $\hat{X}$.

Problems

1. Two observables $\hat{A}$ and $\hat{B}$ are represented in an orthonormal basis by

$$\left[\hat{A} \right] = \begin{bmatrix} 1 & 0 & 0 \\ 0 & 3 & 0 \\ 0 & 0 & 3 \end{bmatrix} \quad \text{and} \quad \left[\hat{B} \right] = \begin{bmatrix} -1 & 0 & 0 \\ 0 & 4 & -5i \\ 0 & 5i & 4 \end{bmatrix}$$

   (a) Argue without computing the commutator explicitly that $\hat{A}$ and $\hat{B}$ are compatible.

   (b) Identify why the form of $\hat{B}$ is block-diagonal.

   (c) Verify that they form a nontrivial CSCO (i.e., none of them is a CSCO by itself).

   (d) Construct the orthonormal basis comprising common eigenvectors of $\hat{A}$ and $\hat{B}$. Label the members of the basis by the eigenvalues of $\hat{A}$ and $\hat{B}$.

   (e) Construct an observable for this system which will form a CSCO by itself[19]

---

[18]Incidentally, the uncertainty relation was first stated for the position and momentum operators $\hat{x}$ and $\hat{p}$ (to be introduced later in this book). The commutator of $\hat{x}$ and $\hat{p}$ is given by what is known as the canonical commutation relation: $[\hat{x}, \hat{p}] = i\hbar$, where $\hbar$ is a fundamental constant of nature called the Planck's constant (also to be introduced later). This gives the original version of the uncertainty relation:

$$\sigma_x \sigma_p \geq \frac{\hbar}{2}$$

[19]You will have thus demonstrated that the number of members in a CSCO is indeed a feature of the description, and not of the system.

2. For the following two pairs of operators, check whether there exists a complete set of common eigenvectors.

$i)$ $\left[\hat{A}\right] = \begin{bmatrix} 3 & 2 \\ 0 & 3 \end{bmatrix}$ and $\left[\hat{B}\right] = \begin{bmatrix} 7 & 0 \\ 0 & 7 \end{bmatrix}$

$ii)$ $\left[\hat{X}\right] = \begin{bmatrix} 3 & 1 \\ 1 & 3 \end{bmatrix}$ and $\left[\hat{Y}\right] = \begin{bmatrix} 2 & 3 \\ 3 & -2 \end{bmatrix}$

3. Investigate if the operators given below are compatible, and whether they form a CSCO.

$[\hat{\alpha}] = \begin{bmatrix} 4 & -4 & 2 \\ -4 & 4 & -2 \\ 2 & -2 & 1 \end{bmatrix}$ and $[\hat{\beta}] = \begin{bmatrix} 5 & 4 & -2 \\ 4 & 5 & 2 \\ -2 & 2 & 8 \end{bmatrix}$

4. Consider the two observables $\hat{G}$ and $\hat{H}$ given in some representation as

$\left[\hat{G}\right] = \begin{bmatrix} 1 & 0 & 1 \\ 0 & 0 & 0 \\ 1 & 0 & 1 \end{bmatrix}$ and $\left[\hat{H}\right] = \begin{bmatrix} 2 & 1 & 1 \\ 1 & 0 & -1 \\ 1 & -1 & 2 \end{bmatrix}$

(a) Show that $\hat{G}$ and $\hat{H}$ represent compatible observables.

(b) The set $\left\{ (1/\sqrt{3})(1,-1,-1)^T, (1/\sqrt{6})(-1,-2,1)^T, (1/\sqrt{2})(1,0,1)^T \right\}$ is an eigenbasis of $\hat{G}$. Construct an orthonormal basis of common eigenvectors of $\hat{G}$ and $\hat{H}$. You are not allowed to solve for the eigenvectors of $\hat{H}$ directly[20].

(Hint: Write down the representation of the *restriction* of $\hat{H}$ to the eigenspace associated with any degenerate eigenvalue of $\hat{G}$ if present. Diagonalize this restriction, and transform back to the original basis.)

5. Prove the results of the last problem of the last problem set in a general context:

(a) Show that the final state resulting from simultaneous measurements of two compatible observables $\hat{A}$ and $\hat{B}$ is independent of the order of the measurement[21]:

$$|\psi; A_i, B_j\rangle = |\psi; B_j, A_i\rangle$$

[20] After you have solved the problem, can you see why you were not allowed to solve the eigenvalue problem of $\hat{H}$ directly?

[21] The wording of this problem and the one to follow brings out the special meaning of "simultaneous measurement" in QM. We say *simultaneous* and yet we talk of *order*.

(b) Using the previous result show that if a series of simultaneous measurements of $\hat{A}$, $\hat{B}$ and $\hat{A}$ are performed (in the given order), the outcome of $\hat{A}$ obtained in the first measurement will be reproduced in the last

$$P\left[\|\psi; A_i, B_j\rangle \to A_k\right] = \delta_{ik}$$

(c) Show that simultaneous measurements of a CSCO $\left\{\hat{A}, \hat{B}, \hat{C}\right\}$ results in a final state that is independent of the initial state.

$$|\psi; A_i, B_j, C_k\rangle = |\phi; A_i, B_j, C_k\rangle = |\chi; A_i, B_j, C_k\rangle = \ldots$$

where $|\psi\rangle, |\phi\rangle, |\chi\rangle, \ldots$, etc., are arbitrary initial states. The final states can then simply be written as[22] $|A_i, B_j, C_k\rangle$.

6. Prove that an alternative definition of compatibility of a collection of observables could be provided using either of the following criteria:

(a) The state resulting from a sequence of instantaneously subsequent measurements of the observables is independent of the order of the measurements.

(b) If the measurement of one of the observables of the collection is repeated soon after a set of simultaneous measurements of observables from the collection, the outcome recorded during the simultaneous measurement will be reproduced.

7. Show that, if $\hat{A}, \hat{B}, \hat{C}$ are linear operators acting on some vector space, and b, c are scalars, then the following commutation properties hold:

(a) $\left[\hat{A}, \hat{B}\right] = -\left[\hat{B}, \hat{A}\right]$

(b) $\left[\hat{A}, b\hat{B} + c\hat{C}\right] = b\left[\hat{A}, \hat{B}\right] + c\left[\hat{A}, \hat{C}\right]$

(c) $\left[\hat{A}, \hat{B}\hat{C}\right] = \left[\hat{A}, \hat{B}\right]\hat{C} + \hat{B}\left[\hat{A}, \hat{C}\right]$

8. Prove the uncertainty relation:

$$\sigma_A^2 \sigma_B^2 \geq \left(\frac{1}{2i}\left\langle\left[\hat{A}, \hat{B}\right]\right\rangle\right)^2$$

where $\hat{A}$ and $\hat{B}$ are two incompatible observables having a commutator $\left[\hat{A}, \hat{B}\right]$ and

$$\sigma_X = \sqrt{\left\langle(X - \langle X\rangle)^2\right\rangle}$$

is the standard deviation of the observable[23] $\hat{X}$.

---

[22] These results are trivially generalized to sets containing more compatible observables.
[23] Can you see why the result is meaningful in spite of the factor i on the right hand side?

Chapter 6

Time Evolution

Up until now, we have been considering instantaneously subsequent measurements. This allowed us to restrict ourselves to the scenario in which measurements were performed in such quick succession that quantum states did not have time to evolve in between measurements. Now we want to look at the more general scenario which allows for *evolution* of the quantum states in *time*. This means we will look into the change in a state that is brought about simply by *waiting* (as opposed to making measurements). But before we delve into that, we will have to introduce some mathematical prerequisites.

Unitary Operators

To discuss time evolution in QM, the main tool that will be needed is a *unitary operator*. But to define a unitary operator we must first introduce the *identity* and the *inverse* operators.

Identity operator

The **identity operator** $\hat{I}$ acting on a vector space V is defined by

$$\hat{I}|\psi\rangle = |\psi\rangle$$

for all vectors $|\psi\rangle$ belonging to V. Clearly, for every linear operator $\hat{A}$ acting on V we have

$$\hat{I}\hat{A} = \hat{A}\hat{I} = \hat{A}$$

The identity operator acting on an n-dimensional vector space is represented by an $n \times n$ identity matrix $\left[\hat{I}\right]$.

Inverse operator

The **inverse $\hat{A}^{-1}$ of an operator** $\hat{A}$ is defined by

$$\hat{A}\hat{A}^{-1} = \hat{A}^{-1}\hat{A} = \hat{I}$$

Not every operator has an inverse. Operators that do have an inverse are called ***invertible operators***. In general for every operator $\hat{A}$ we have

$$\hat{A} \, | \, \rangle = | \, \rangle$$

where $| \, \rangle$ is the additive identity vector. If there exists a vector $|\psi\rangle \neq | \, \rangle$ such that

$$\hat{A} \, |\psi\rangle = | \, \rangle$$

then $\hat{A}$ is said to be a ***singular operator***, and in this case $|\psi\rangle$ is called the ***singular point***. An operator which is not singular is called a ***nonsingular operator***. It turns out that *an operator is invertible if and only if it is nonsingular*. Incidentally, an operator $\hat{A}$ is nonsingular if and only if its representation $\left[\hat{A}\right]$ has a *nonvanishing determinant*. Such a matrix is also called nonsingular[1].

If an operator $\hat{A}$ is represented by the matrix $\left[\hat{A}\right]$, its inverse is represented by the inverse matrix $\left[\hat{A}\right]^{-1}$.

Unitary operators

An operator $\hat{U}$ is said to be a ***unitary operator*** if

$$\hat{U}^{-1} = \hat{U}^{\dagger}$$

That is, if

$$\hat{U}\hat{U}^{\dagger} = \hat{U}^{\dagger}\hat{U} = \hat{I}$$

An unitary operator $\hat{U}$ is represented by an unitary matrix $\left[\hat{U}\right]$:

$$\left[\hat{U}\right]\left[\hat{U}\right]^{\dagger} = \left[\hat{U}\right]^{\dagger}\left[\hat{U}\right] = \left[\hat{I}\right]$$

Important properties of unitary operators

The following, easy to prove, properties of unitary operators are important in the analysis of time evolution in QM:

- *A necessary and sufficient condition for an operator to be unitary is that it preserves the norm of every vector that it acts on*[2].

- *An unitary operator can always be written as the exponential of an anti-Hermitian operator*[3].

---

[1]You should be able to prove these results right away. To prove the second result, one needs to make use of the fact that the columns (or rows) of a square matrix are linearly independent if and only if it has a nonvanishing determinant.

[2]It is trivial to see why the condition of unitarity is *sufficient* to preserve norm. That the condition is necessary is also not difficult to show. You will be asked to prove this in the problem set at the end of this chapter.

[3]You will prove this also in the problem set. An anti-Hermitian operator, by the way, can always be written as 'i' times a Hermitian operator.

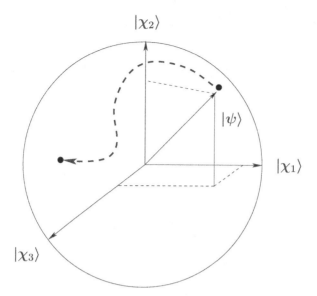

$|\chi_2\rangle$

$|\psi\rangle$

$|\chi_1\rangle$

$|\chi_3\rangle$

FIGURE 6.1: Time Evolution. The trajectory of the state vector of a 3-dimensional quantum system lies on the surface of a unit sphere embedded in 3-dimensional space. Here we have, yet again, indulged in the oversimplification of using a real vector space for the purpose of visualization.

Time Evolution of Quantum States

Let us consider a state $|\psi(t)\rangle$ at time t that has evolved from the state $|\psi(t_0)\rangle$ at some earlier time t_0 (see Figure 6.1). Our purpose in this section is to look for a clue which would provide a *law* that would enable us to determine $|\psi(t)\rangle$ from $|\psi(t_0)\rangle$.

Time evolution operator for conservative systems

It is perhaps only too natural that one would explore the possibility that the state $|\psi(t)\rangle$ is linearly related to $|\psi(t_0)\rangle$. So, we want to look into the possibility that there exists an operator $\hat{U}(t, t_0)$ such that[4]

$$|\psi(t)\rangle = \hat{U}(t, t_0) |\psi(t_0)\rangle$$

---

[4]We are trying to look into the possibility that when expressed in a representation (say χ), every component of the vector $|\psi(t)\rangle$ would be a linear function (i.e., a linear combination) of the components of the vector $|\psi(t_0)\rangle$. This can be written as a matrix equation

$$[\psi(t)]^\chi = U(t, t_0) [\psi(t_0)]^\chi$$

Although we have written $\hat{U}(t, t_0)$ as a function of two arguments t and t_0, it is to be understood that it depends only on the difference $t - t_0$. For this equation to be acceptable, it must ensure that an initial quantum state evolves into a vector that is also a quantum state. That is, the equation must preserve the normalization of the initial state. Recall, that the normalization condition on a state $|\psi\rangle$ is expressed as

$$\langle \psi | \psi \rangle = 1$$

Now, if we start from the normalized state $|\psi(t_0)\rangle$ and require that the final state $|\psi(t)\rangle$ is also normalized, then

$$\langle \psi(t) | \psi(t) \rangle = 1$$
$$\implies \quad \langle \psi(t_0) | \, \hat{U}^\dagger(t, t_0) \hat{U}(t, t_0) \, | \psi(t_0) \rangle = 1$$

Now we know that this requires that the operator $\hat{U}(t, t_0)$ must be unitary. Hence

$$\hat{U}^\dagger(t, t_0) \hat{U}(t, t_0) \quad = \quad \hat{I}$$

Of course, finally, it can only be decided by experiment whether or not the time evolution of states may actually be described by an unitary, linear operator as laid out above. Indeed, such a description has been validated overwhelmingly by experiment. The unitary operator $\hat{U}(t, t_0)$ that governs the time evolution of quantum states is called the **time evolution operator** (sometimes, simply called *evolution operator*).

Now, since one can always write a unitary operator as an exponential of an anti-Hermitian operator, or equivalently, the exponential of i times a Hermitian operator[5], one can write the evolution operator as

$$\hat{U}(t, t_0) = \exp\left\{ i \hat{\mathcal{H}} \right\}$$

Here $\hat{\mathcal{H}}$ is a function of the duration $t - t_0$.

In the preceding discussion we started by exploring the simple scenario that the final state has a linear dependence on the initial state. Now we wish to make a further simplification. We consider a situation where the *temporal* dependence of the Hermitian operator $\hat{\mathcal{H}}$ is also linear[6]. This means that the duration $t - t_0$ occurs as a multiplicative factor in $\hat{\mathcal{H}}$. It turns out that for a large class of systems, we can indeed choose to write the *law of time evolution* in this form. Such systems are called *conservative*. It is customary to write

---

where $U(t, t_0)$ is some square matrix of appropriate dimension. In the abstract vector space, this will manifest itself as

$$|\psi(t)\rangle = \hat{U}(t, t_0) \, |\psi(t_0)\rangle$$

for some linear operator $\hat{U}(t, t_0)$.

[5]Physicists usually prefer to use a Hermitian operator.

[6]Do you think that the simpler possibility that $\hat{U}(t, t_0)$ is linear in $t - t_0$ is plausible?

the evolution operator as

$$\hat{U}(t, t_0) = \exp\left\{-\frac{i}{\hbar}\hat{H}(t - t_0)\right\}$$

where $\hat{H}$ is again a Hermitian operator called the **Hamiltonian** of the system. It is assumed to be time independent here so that the time dependence in the exponent enters simply as a multiplicative factor $t - t_0$. The $\hbar$ appearing in the denominator of the exponent is an universal constant called the *Planck's constant* and the negative sign is a convention[7]. It is then clear that

- *the entire information about the dynamics of the quantum mechanical system is contained in the Hamiltonian operator $\hat{H}$.*

Owing to its Hermiticity, $\hat{H}$ can be regarded as an observable[8].

Schrödinger equation

Now we shall try to arrive at a differential form of the law of time evolution. Let us imagine that a state $|\psi(t)\rangle$, of a conservative system, evolves into a state $|\psi(t + \delta t)\rangle$ in a time interval δt. Using the time evolution equation, we can write

$$|\psi(t + \delta t)\rangle = \hat{U}(\delta t)|\psi(t)\rangle$$

$$\text{with} \quad \hat{U}(\delta t) = \exp\left\{-\frac{i}{\hbar}\hat{H}\,\delta t\right\}$$

Expanding out the exponential, we have

$$|\psi(t + \delta t)\rangle = \left(\hat{I} - \frac{i}{\hbar}\hat{H}\,\delta t + \ldots\right)|\psi(t)\rangle$$

$$\frac{\psi(t + \delta t) - \psi(t)}{\delta t} = -\frac{i}{\hbar}\hat{H}|\psi(t)\rangle + \ldots$$

---

[7]In 1900, in an attempt to find the elusive description of the intensity of electromagnetic radiation from a black-body as a function of the frequency of radiation, the German physicist Max Planck hypothesized that the energy from radiation was available only in discrete bundles (quanta) and that the minimum quantum of energy E_ν was proportional to the frequency of the radiation ν. The relationship was written as $E = h\nu$ and the constant h was subsequently called the Planck's constant. This marked the birth of QM. For theoretical convenience $\hbar = h/2\pi$ was defined later. It turned out that it is the quantum transcription of energy (Hamiltonian) that governs the time evolution of systems. That is how the constant crept into this equation.

[8]To know how $\hat{H}$ got to be called the Hamiltonian, once again, one has to go into the history of the development of QM. Unfortunately, we will be unable to discuss it in this book. However, I urge the reader to look up the literature to know about this intriguing story.

where in the last equation, the dots contain terms involving δt. Taking the limit $\delta t \to 0$ gives the differential equation:

$$i\hbar \frac{d\,|\psi(t)\rangle}{dt} = \hat{H}\,|\psi(t)\rangle$$

Although, the above equation was derived for conservative systems, it turns out that this differential form of the time evolution equation is actually *more general*! It governs the time evolution of *all* quantum systems (and not just the conservative systems from which it was derived). It is called the **Schrödinger equation** after the Austrian physist Erwin Scröodinger who discovered it.

The Law of Time Evolution *The Schrödinger equation embodies the general law of time evolution of states in QM.*

We have derived the Schrödinger equation from the unitary state evolution formula. One can also show the converse that unitary evolution follows from the Schrödinger equation. If the Hamiltonian does not depend explicitly on time, then it is trivial to see that the Schrödinger equation leads to a time evolution operator given precisely by the form $\hat{U}(t) = \exp\left\{-(i/\hbar)\,\hat{H}\,t\right\}$. For Hamiltonians that have an explicit time dependence, the time evolution operator is quite a bit more involved. Such systems are called *nonconservative*. We shall be mostly concerned with conservative systems in this book.

Postulates of Quantum Mechanics - Version 4

The law of time evolution of states will obviously have to go into our list of postulates of QM. For the sake of completeness we will write down the full list of postulates one more time by including the time evolution postulate.

1. Every system is associated with an inner product space. States of the system are the normalized vectors belonging to this space.

2. Every observable of the system is associated with a linear Hermitian operator acting on the state space.

3. The eigenvalues of the Hermitian operator associated with an observable constitute the measurement spectrum of the observable.

4. If a measurement of an observable $\hat{\phi}$ is made on some state $|\psi\rangle$ that yields an eigenvalue ϕ_j, soon after the measurement, the new state becomes $|\psi\,;\,\phi_j\rangle$ which is given by the *collapse postulate*:

$$|\psi\,;\,\phi_j\rangle = \frac{\sum_{r=1}^{g_j}|\phi_j^r\rangle\langle\phi_j^r|\psi\rangle}{\sqrt{\sum_{r=1}^{g_j}\left|\langle\phi_j^r|\psi\rangle\right|^2}}$$

where $\left|\phi_j^r\right\rangle$ are orthonormal eigenvectors of the eigenvalue ϕ_j and g_j is its degree of degeneracy.

5. The probability $P\left[|\psi\rangle \to |\phi\rangle\right]$ of an event $|\psi\rangle \to |\phi\rangle$ is given by the *Born rule*:

$$P\left[|\psi\rangle \to |\phi\rangle\right] \quad = \quad |\langle\phi|\psi\rangle|^2$$

where the scalar product $\langle\phi|\psi\rangle$ is called the probability amplitude of the event $|\psi\rangle \to |\phi\rangle$.

6. The time evolution of a state in QM is governed by the *Schrödinger equation*:

$$i\hbar\frac{d\,|\psi(t)\rangle}{dt} = \hat{H}\,|\psi(t)\rangle$$

where $\hat{H}$ is a linear, Hermitian operator (called the Hamiltonian).

The Programme in Quantum Mechanics

Finally, we are in a position to lay out a *prescription* to solve the target questions in QM. Let us recall, one final time, the target questions in QM in its full glory (see Figure 6.2):

Given a state $[\psi(t_0)]^\chi$ at some initial time t_0, in some representation χ, if we choose to make a measurement of some observable $\hat{\phi}$ at some later time t, then what are the possible outcomes and what are their probabilities?

The solution to these questions will involve two essential steps:

1. Solution of the Schrödinger's equation, leading to the determination of the state $[\psi(t)]^\chi$ at time t.

2. Solution of the eigenvalue problem for the observable $\hat{\phi}$ yielding the spectrum $\{\phi_i\}$ of $\hat{\phi}$ obtained as the eigenvalues of $\left[\hat{\phi}\right]^\chi$, and the corresponding states $[\phi_i]^\chi$ obtained as the normalized eigenvectors of $\left[\hat{\phi}\right]^\chi$.

The probabilities, $P\left[\psi(t) \to \phi_i\right]$ are then trivially given by

$$P\left[\psi(t) \to \phi_i\right] \quad = \quad |\langle\psi(t); \phi_i|\psi(t)\rangle|^2$$
$$\text{with} \qquad \langle\psi(t); \phi_i|\psi(t)\rangle \quad = \quad \left([\psi(t); \phi_i]^\chi\right)^\dagger [\psi(t)]^\chi$$

This constitutes the essential recipe for solving the basic problem of QM.

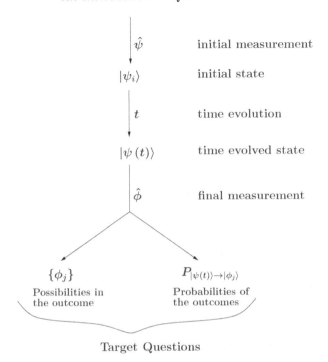

Target Questions

FIGURE 6.2: The full setting of a quantum mechanical problem and the "Target Questions".

Formal Solution of Time Evolution

As mentioned, we will restrict our discussion to time independent Hamiltonians. Since the Hamiltonian is a Hermitian operator, we will be able to construct an orthonormal basis out of its eigenvectors. Let us assume, for simplicity, that the eigenvalues of the Hamiltonian are nondegenerate[9]. Let the eigenvalues of our Hamiltonian $\hat{H}$ be denoted by E_k with k running over its entire spectrum. We can expand a state $|\psi(t)\rangle$ in the orthonormal basis of eigenvectors of the Hamiltonian as

$$|\psi(t)\rangle \;=\; \sum_k |E_k\rangle\,\langle E_k|\psi(t)\rangle$$

---

[9]Nothing in the argument depends on this assumption. It just makes our life easier by allowing for a simpler notation.

By plugging this expansion into Schrödinger's equation we have[10]

$$i\hbar\frac{d}{dt}\left(\sum_k |E_k\rangle\langle E_k|\psi(t)\rangle\right) = \hat{H}\left(\sum_k |E_k\rangle\langle E_k|\psi(t)\rangle\right)$$

$$\sum_k |E_k\rangle\left(i\hbar\frac{d}{dt}\langle E_k|\psi(t)\rangle\right) = \sum_k E_k |E_k\rangle\langle E_k|\psi(t)\rangle$$

Comparing the coefficients of $|E_k\rangle$ we have

$$\frac{d}{dt}\langle E_k|\psi(t)\rangle = -\frac{i}{\hbar}E_k\langle E_k|\psi(t)\rangle$$

which admits of a simple solution[11]

$$\langle E_k|\psi(t)\rangle = \langle E_k|\psi(0)\rangle\, e^{-\frac{i}{\hbar}E_k t}$$

Hence the solution to Schrödinger's equation becomes

$$|\psi(t)\rangle = \sum_k |E_k\rangle\langle E_k|\psi(0)\rangle\, e^{-\frac{i}{\hbar}E_k t}$$

The above exercise shows us that if we know the eigenvalues and eigenvectors of the Hamiltonian, the Schrödinger equation is, in effect, solved. We had seen earlier that for instantaneously subsequent measurements, the target questions of QM were reduced to solving an eigenvalue problem. When we allow time evolution, in the case of conservative systems, we see that the only additional task, that of solving the Schrödinger's equation, also turns out to be equivalent to solving an eigenvalue problem. This time, it is the eigenvalue problem of the Hamiltonian.

Incidentally, it is trivial to see that that the eigenstates of the Hamiltonian do not evolve in time (in any physically significant way). These states are called *stationary states* and are of paramount importance in QM.

Problems

1. The Hamiltonian for a system, represented in the χ space is given by

$$\left[\hat{H}\right]^\chi = \varepsilon\begin{bmatrix} 3 & 2 \\ 2 & 3 \end{bmatrix}$$

---

[10]Here we use the fact that the eigenvectors of the Hamiltonian are time independent. This is obvious since the Hamiltonian has been assumed to be time independent.

[11]This solution can also be simply derived by making the time evolution operator $\hat{U}(t) = \exp\left\{-(i/\hbar)\hat{H}t\right\}$ act on the state $|\psi(t)\rangle = \sum_k |E_k\rangle\langle E_k|\psi(t)\rangle$.

(ε is some dimensionful constant). If the system starts out from an initial state $[\psi(0)]^\chi$ given by

$$i) \quad \begin{bmatrix} 1 \\ 0 \end{bmatrix} \qquad\qquad ii) \quad \frac{1}{\sqrt{2}} \begin{bmatrix} 1 \\ 1 \end{bmatrix}$$

and one makes a measurement of the observable $\hat{A}$ represented in the χ space by

$$\begin{bmatrix} \hat{A} \end{bmatrix}^\chi = \hbar \begin{bmatrix} 6 & 2+3i \\ 2-3i & -6 \end{bmatrix}$$

after t seconds, find out

(a) the different possibilities in the outcome, and

(b) the probabilities of the different possibilities.

2. Consider the Hamiltonian of a 4-level system:

$$\begin{bmatrix} \hat{H} \end{bmatrix} = E \begin{bmatrix} 3 & 0 & 2 & 0 \\ 0 & 6 & 0 & 2+3i \\ 2 & 0 & 3 & 0 \\ 0 & 2-3i & 0 & -6 \end{bmatrix}$$

where E is a dimensionful constant.

(a) Find the state at time t if the system starts out from the initial state

$$i) \quad \frac{1}{\sqrt{2}} \begin{bmatrix} 1 \\ 0 \\ -1 \\ 0 \end{bmatrix} \qquad\qquad ii) \quad \begin{bmatrix} 0 \\ 0 \\ 0 \\ 1 \end{bmatrix}$$

(Hint: Do not use brute force. You should be able write down the solution right away from the previous problem.)

(b) Find the probability that a measurement outcome of $\hat{H}$ will yield the value E after 10 billion years in each case.

3. At $t = 0$ a measurement of an observable $\hat{B}$ is made on a 4-level system which yields the value $-b$. The system then evolves under a Hamiltonian $\hat{H}$. The representations of $\hat{B}$ and $\hat{H}$ are given in some orthonormal basis by

$$\begin{bmatrix} \hat{B} \end{bmatrix} = \begin{bmatrix} 0 & b & 0 & 0 \\ b & 0 & 0 & 0 \\ 0 & 0 & 2b & 0 \\ 0 & 0 & 0 & 3b \end{bmatrix} \quad \text{and} \quad \begin{bmatrix} \hat{H} \end{bmatrix} = \begin{bmatrix} \varepsilon & 0 & 0 & 0 \\ 0 & 2\varepsilon & 0 & 0 \\ 0 & 0 & 2\varepsilon & 0 \\ 0 & 0 & 0 & 3\varepsilon \end{bmatrix}$$

After time t, find out

   (a) the probability that a measurement of $\hat{B}$ yields $-b$ again, and

   (b) the probability that a measurement of $\hat{H}$ yields 2ε.

4. Show that the Schrödinger equation preserves the normalization of states.

5. Prove that, if an operator $\hat{A}$ satisfies

$$\left\langle \psi | \hat{A}^\dagger \hat{A} | \psi \right\rangle = 1$$

then it must be unitary.

6. Show that an unitary matrix can always be expressed as the exponential of an anti-Hermitian matrix[12].

7. Show that for conservative systems, the Schrödinger equation implies that the evolution operator $\hat{U}(t - t_0)$ is given by

$$\hat{U}(t - t_0) = \exp\left\{ -\frac{i}{\hbar} \hat{H}(t - t_0) \right\}$$

where $\hat{H}$ is the Hamiltonian operator, and $t - t_0$ is the temporal interval. (Assume that the standard rules of calculus can be used for operators.)

8. For a nonconservative system, the Hamiltonian $\hat{H}(t)$ depends explicitly on time. From the Schrödinger equation, the evolution operator may appear to be

$$\hat{U}(t, t_0) = \exp\left\{ -\frac{i}{\hbar} \int_{t_0}^{t} \hat{H}(t') \, dt' \right\}$$

This is, however, incorrect. Figure out why.

9. Suppose the Hamiltonian $\hat{H}$ of a certain system can be written as $\hat{H} = \hat{H}_0 + \hat{H}_1$. Assume further that $\hat{H}_1$ commutes with $\hat{H}_0$. If we have the system in an initial state $|E_0\rangle$, which is an eigenstate of $\hat{H}_0$, write down a formal expression for the probability that at a later time the system may be found in the state $|E_0\rangle$.

10. Show that the time evolution of the expectation value of an observable $\hat{A}$ (which can possibly have an explicit time dependence) obeys the relation

$$i\hbar \frac{d\left\langle \hat{A} \right\rangle}{dt} = \left\langle \left[\hat{A}, \hat{H} \right] \right\rangle + i\hbar \frac{\partial \left\langle \hat{A} \right\rangle}{\partial t}$$

where $\hat{H}$ is the Hamiltonian operator.

---

[12]You might need to first establish that unitary operators are necessarily diagonalizable. See appendix 'B' if you are stuck.

11. An observable which does not explicitly depend on time, and also commutes with the Hamiltonian is called a *constant of motion*.

    (a) Show that the probability of getting a measurement outcome A_i of a constant of motion $\hat{A}$ is constant in time.

    (b) Show that an eigenstate of a constant of motion will never evolve into a state that does not belong to the same eigenspace.

    (c) Show that the expectation value of a constant of motion does not evolve over time.

    (d) Constants of motion play an important role in QM. Can you imagine why?

12. An alternative description of time evolution is given by the so called *Heisenberg picture*. In this description all vectors and operators are transformed by a time dependent unitary transformation as follows:

$$\begin{aligned} |\psi\rangle_H &= \hat{U}^\dagger(t) |\psi\rangle_S \\ \hat{A}_H &= \hat{U}^\dagger(t) \hat{A}_S \hat{U}(t) \end{aligned}$$

Here, the subscript H denotes "Heisenberg picture" and the subscript S indicates the usual description, generally referred to as the "Schrödinger picture". The operator $\hat{U}(t)$ is the evolution operator $\hat{U}(t) \equiv \hat{U}(t, 0)$.

    (a) Show that all predictions of QM (i.e., the spectra of observables, and the probabilities) remain unchanged in the Heisenberg picture.

    (b) Observe that, in the Heisenberg picture, states do not evolve in time while observables in general do. Justify the comment: "In the Heisenberg picture, although states are constant in time, the *meaning* of states are not".

    (c) Obtain a formula to describe how the eigenvector of an observable will evolve in time in the Heisenberg picture.

    (d) Show that, in the Heisenberg picture, an operator $\hat{A}_H$ evolves in time according to the differential equation

$$i\hbar \frac{d\hat{A}_H}{dt} = \left[\hat{A}_H, \hat{H} \right] + i\hbar \left[\frac{\partial \hat{A}_H}{\partial t} \right]$$

where the subscript H indicates the Heisenberg picture. Can you see why the Hamiltonian observable $\hat{H}$ does not have the subscript H?

Chapter 7

Continuous Spectra

So far, we have only been considering observables whose spectra are discrete. This was not because continuous spectra are forbidden by nature[1]. It was because the ideas that we were trying to communicate are easier to grasp for discrete spectra. In this chapter we want to discuss observables whose spectra form a continuum. It will, unfortunately, not be possible to discuss this with any mathematical rigor since it will take us way beyond the scope of the present book. It would require formulations involving infinite dimensional inner product spaces and functional analysis. However, the structure of QM in such a scenario closely resembles the quantum mechanical descriptions of systems that involve only finite dimensional inner product spaces which we have been studying so far. This is what we shall exploit. So, the content of this chapter should be looked upon as essentially a collection of results motivated mainly by analogy[2].

This chapter will be of considerable practical importance since most real life quantum systems will use descriptions that are discussed here.

Description using Function Spaces

We will now quickly re-run the developments leading to the inner product space formulation of QM[3]. Only this time, we will start from a scenario where we contemplate states to be described in some χ space, where $\hat{\chi}$ is an

---

[1] Well, in a sense they are, but that is not why we chose to avoid them thus far. Notwithstanding the prohibition, such operators play an essential role in the description of many (if not most) quantum systems, as we shall see in this chapter.

[2] Having said that one must also be alert to the differences between descriptions involving discrete, finite dimensional inner product spaces and infinite dimensional inner product spaces required for including observables with continuous spectra. These differences are often stark, counter-intuitive and disconcerting. For the puritan, let us admit that many of the statements that we shall make in this chapter are not even strictly *correct* from a mathematical standpoint. Our presentation will be purely heuristic. The justification for this naive approach is that if the scheme outlined here is followed, it leads to the *correct* answers for most standard applications of QM.

[3] We are, in effect, re-starting our story from the chapter "States as Vectors".

observable that has a continuous spectrum[4]. Now, if we attempt to use such a space for our description, we shall see that we will need to make some changes in our language, notation and interpretation[5]. Let us see what these changes might be.

Vectors

If the spectrum of the observable $\hat{\chi}$ was enumerable and finite (the case that we have seen so far), we would have written the spectrum as a set $\{\chi_i\,; i = 1, 2, \ldots n\}$ (where n denotes the cardinality of the set) so that a vector $|\psi\rangle$ would be represented in the χ space by an ordered list of n complex numbers written as a column matrix

$$[\psi]^\chi = \begin{bmatrix} \langle\chi_1|\psi\rangle \\ \langle\chi_2|\psi\rangle \\ .. \\ .. \\ \langle\chi_n|\psi\rangle \end{bmatrix}$$

We have seen that the representations $[\psi]^\chi$ are also considered to be vectors, but this time in the representation space $\mathbb{C}^n$. For such a vector $[\psi]^\chi$, we have one complex number $\langle\chi_i|\psi\rangle$ (called a component) corresponding to each member χ_i of the spectrum of $\hat{\chi}$. Now, let us try to transcribe the object $[\psi]^\chi$ for the setting where the spectrum of $\hat{\chi}$ is continuous. Let us imagine that the spectrum of $\hat{\chi}$ is the real interval $[a, b]$. We shall denote the members of the spectrum by the symbol χ. Then the transcribed object, which we tentatively *call* a vector as well, should have a component for each real number in $[a, b]$. But this, by definition, is a *complex function defined over the domain* $[a, b]$. So,

- *a vector in the χ space would now be a function $\psi(\chi)$ instead of a column matrix $[\psi]^\chi$:*

$$[\psi]^\chi \longrightarrow \psi(\chi)$$

The value of the function for a particular value χ of the domain would be a *component* of the vector, and will be denoted by $\psi(\chi) \equiv \langle\chi|\psi\rangle$ in accordance with the notation $\langle\chi_i|\psi\rangle$ for the enumerable case (see Figure 7.1). Of course, whether such vectors actually form a vector space or not must be checked by testing whether such functions obey the defining properties of vector spaces under addition and multiplication by a scalar. This is rather easy to do and one can easily convince oneself that the set of complex functions over the real

---

[4]As we had done before, we are also making the tacit assumption that we are in a scenario where $\hat{\chi}$ forms a CSCO by itself so that there is no degeneracy, etc. The description we demonstrate below will easily carry over to situations when one needs more members to complete the CSCO.

[5]We will also see to what extent it is actually possible to regard $\hat{\chi}$ as a *true* observable.

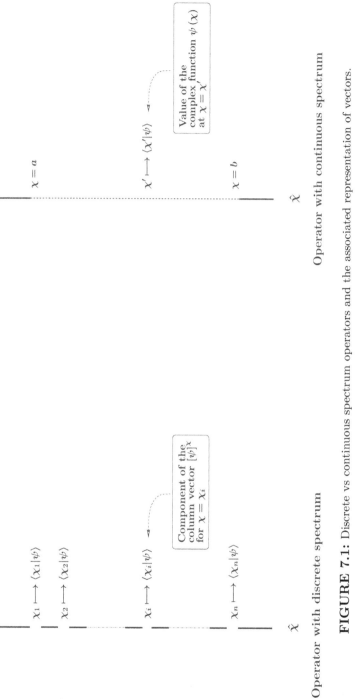

FIGURE 7.1: Discrete vs continuous spectrum operators and the associated representation of vectors.

interval $[a, b]$ is indeed a vector space under addition of complex functions and multiplications of such functions by a complex number. Let us call this set $\mathbb{F}^c[a, b]$.

Next, we would like to investigate whether we can identify the standard basis in $\mathbb{F}^c[a, b]$. Recall that the column matrices $\{[\chi_k] ; k = 1, 2, \ldots, n\}$ whose components are defined by the kronecker delta function $\langle \chi_i | \chi_k \rangle = \delta_k^i$ provided the standard basis in the discrete case:

$$\delta_k^i = \begin{cases} 0 & \text{if} \quad \chi_k \neq \chi_i \\ 1 & \text{if} \quad \chi_k = \chi_i \end{cases}$$

$$\psi^i = \sum_{k=1}^{n} \delta_k^i \psi^k$$

$$[\psi]^\chi = \sum_{k=1}^{n} [\chi_k]^\chi \psi^k$$

where the index k runs from 1 to n. In the continuum, the so called **Dirac-delta functions** $\delta (\chi - \chi')$, defined by

$$\delta (\chi - \chi') = \begin{cases} 0 & \text{if} \quad \chi \neq \chi' \\ \infty & \text{if} \quad \chi = \chi' \end{cases}$$

$$\psi (\chi') = \int_a^b d\chi \, \delta (\chi - \chi') \, \psi (\chi)$$

provide a *standard basis*. That is, the standard basis in the continuum are a set of functions $\{\chi (\chi') ; \chi' \in [a, b]\}$ such that the components $\langle \chi' | \chi \rangle$ of the functions $\chi (\chi')$ are given by

$$\langle \chi' | \chi \rangle = \delta (\chi - \chi')$$

The definition of the Dirac-delta function directly ensures that they qualify as the standard basis in an analogous fashion to the discrete case cited above[6].

Inner products

We can attempt to endow the vector space of all complex functions over $[a, b]$ with an inner product defined by

$$\langle \phi | \psi \rangle = \int_a^b d\chi \, \phi (\chi)^* \, \psi (\chi)$$

---

[6]The Dirac-delta functions are, technically, not functions. In proper mathematical terms, they belong to what are known as *distributions* (in fact, they are examples of what are known as a *non-regular* distributions). This means that the basis functions cannot belong to our vector space $\mathbb{F}^c[a, b]$! This is extremely disconcerting and certainly needs to be legitimized. Unfortunately, it is beyond the scope of this book. The interested reader is strongly encouraged to look into the meaning and justification more carefully at some stage. For this, a comprehensive study of the formulation of QM on infinite dimensional inner product spaces which admit observables with continuous spectrum must be taken up.

in analogy with the corresponding formula for the finite, enumerable case. Here $\psi(\chi)$ and $\phi(\chi)$ are two vectors in $\mathbb{F}^c[a,b]$.

Even before we try to check whether the above formula conforms to the defining properties of inner products we must, of course, ensure that the above integral would converge. Now, in order that this integral converges for *all* functions $\phi(\chi)$, it is clearly *necessary* (in order that the above equation holds for $\phi = \psi$) that

$$\langle\psi|\psi\rangle = \int_a^b d\chi\,\psi(\chi)^*\,\psi(\chi) = \int_a^b d\chi\,|\psi(\chi)|^2$$

should converge. Such functions are called **square integrable functions**. It is quite remarkable that if we assume that two functions $\psi(\chi)$ and $\phi(\chi)$ are both square integrable, then the condition of square integrability turns out to be *sufficient* for the integral $\int_a^b d\chi\,\phi(\chi)^*\,\psi(\chi)$ to converge. One can easily check that

- *in the set of square integrable functions, the proposed definition of inner product holds obeying all the required properties of inner products.*

Thus,

- *in the continuum, the state space of quantum systems must be the space of square integrable functions[7].*

The space of square integrable functions defined over $[a,b]$ is denoted by $\mathbb{L}^2[a,b]$. Another such space that is used commonly in QM is $\mathbb{L}^2(-\infty,+\infty)$, the meaning of which is self evident. Technically, these spaces are called **Hilbert Spaces**[8].

We would naturally define the **norm** $|\psi|$ of a vector $|\psi\rangle$ by

$$|\psi| = \sqrt{\langle\psi|\psi\rangle}$$

In $\mathbb{L}^2[a,b]$ all vectors can be normalized so that

$$\int_a^b d\chi\,\psi(\chi)^*\,\psi(\chi) = \int_a^b d\chi\,|\psi(\chi)|^2 = 1$$

---

[7]It is trivial to check that the space of square integrable functions is also *closed* under addition and multiplication by a scalar. Therefore such spaces will automatically obey the defining conditions of a vector space.

[8]Strictly speaking, these are only two examples of Hilbert Spaces. You should look up the precise definition at some point. It turns out that all finite dimensional inner product spaces trivially qualify as Hilbert spaces. But Hilbert spaces really come to there own for the infinite dimensional cases. Unfortunately, it turns out that even infinite dimensional Hilbert spaces are not adequate when we wish to describe QM with observables that have a continuous spectrum! The appropriate mathematical setting in this case is a, so called, *rigged Hilbert space*.

If we compute the inner product between the standard basis elements, we have

$$\langle \chi' | \chi \rangle = \delta (\chi - \chi')$$

We *define* this to be the new **orthonormality condition** for continuous bases. This is obviously not the kind of orthonormality that we have seen in the enumerable case. However, this definition of orthonormality correctly reproduces the desired relations (that are analogous to the enumerable case). For example, this ensures that the inner product of a vector with a basis element (of an orthonormal basis in the modified sense) correctly picks out the expansion coefficient corresponding to that basis element when the given vector is expanded in the basis.

Identification of the laws

Along the lines that we had followed for the finite, enumerable spectrum scenario, we make the following associations again:

- *Normalized vectors in the space of square integrable functions are quantum states.*

- *Inner products in the space of square integrable functions between state vectors are probability amplitudes[9].*

- *Orthonormal bases in the space of square integrable functions are associated with observables such that every member of such a basis corresponds to an element of the spectrum of the associated observable.*

Here we would like to remark that unlike the usual orthonormal bases, the bases which are orthonormal in the extended sense cannot be associated with *physical* quantum observables[10]. This is because the elements of such a basis cannot really qualify as quantum states. For example, if we look at the standard basis that we have been using, we see that $\langle \chi | \chi \rangle$ is infinite! These vectors are therefore not normalizable (which means that they do not belong to $\mathbb{L}^2$) and do not qualify as quantum states[11].

Before we move on to the next section we would like to mention that for historical reasons, the state vectors when represented in some basis are often called **wave functions**. In fact, this terminology is freely used even in the enumerable basis case.

---

[9]The interpretation of the inner product $\langle \chi | \psi \rangle$ of a square integrable function $\psi (\chi')$ and a continuous basis function such as $\chi (\chi')$ is however slightly different. We shall come to this shortly.

[10]It is, nevertheless, customary to call them observables and we will continue to do so.

[11]Actually, they are not even functions as we have mentioned before. However, they are still useful because, in a sense, they provide a basis for $\mathbb{L}^2$. That turns out to be adequate for our purpose. You have every right to feel confused about this at this stage. Later on we will provide some concrete examples that will demonstrate how one works with such bases. This will, hopefully, make you somewhat more comfortable.

Abstraction

One can imagine an abstract vector space V having a basis $\{|\chi\rangle \; ; \; \chi \in [a,b]\}$, the elements of which are labelled by the continuous string of real numbers $\chi \in [a,b]$, such that a vector $|\psi\rangle$ can be formally written as[12]

$$|\psi\rangle \;=\; \int_a^b d\chi \, |\chi\rangle \langle \chi|\psi\rangle$$

The representation of the vector $|\psi\rangle$ in the basis $\{|\chi\rangle \; ; \; \chi \in [a,b]\}$ is essentially the function $\psi(\chi) = \langle \chi|\psi\rangle$ defined over the real interval $[a,b]$. Restricting the representations $\psi(\chi)$ to square integrable functions defines a subspace of V that can be endowed with an inner product, making it an inner product space.

Incidentally, if we assume that the basis $\{|\chi\rangle \; ; \; \chi \in [a,b]\}$ is orthonormal in accordance with the definition of orthonormality that we have provided in a preceding subsection, then it reproduces the defining expression of inner product in $\mathbb{L}^2[a,b]$ that we had provided earlier. This only requires a straightforward application of linearity and conjugate symmetry property of inner products:

$$
\begin{aligned}
\langle \phi|\psi\rangle \;&=\; \int_a^b \int_a^b d\chi d\chi' \, \phi(\chi)^* \, \psi(\chi') \langle \chi|\chi'\rangle \\
&=\; \int_a^b \int_a^b d\chi d\chi' \, \phi(\chi)^* \, \psi(\chi') \, \delta(\chi' - \chi) \\
&=\; \int_a^b d\chi \, \phi(\chi)^* \, \psi(\chi)
\end{aligned}
$$

In the abstract setting, this should be looked upon as the formula for the inner product of two vectors expressed in terms of their coordinates in an orthonormal basis.

Operators

It is easy to imagine along the lines that was followed for the observables with discrete spectrum that one may associate with the quantum observables having continuous spectrum, linear Hermitian operators with the additional property that their eigenvectors must form a complete set[13].

---

[12]Note that, here $d\chi$ actually stands for the measure of the integral which is the appropriate differential volume in the χ space. For example, if χ refers to the position of a particle moving on a circular track of radius a then $d\chi = ad\theta$ where θ is the polar angle.

[13]Recall that we had mentioned, way back in the fourth chapter that in infinite dimensional inner product spaces, it is not in general guaranteed that the eigenvectors of a Hermitian operator will have a complete set of eigenvectors. Further, we had also mentioned that Hermitian operators that do have a complete set are called "observables", which is a mathematical term in this context.

In the present scenario the representations of such operators will clearly be different. When we are working with a continuous basis, the representations of vectors are functions. Therefore, the *corresponding* representations of linear operators will not be matrices any more. As we shall see in the next section, in the cases of interest to us,

- *the representations of operators will be differential operators*[14].

To state this more concretely, let us use the inner product space V that was introduced in the preceding subsection. We shall denote the representation of an operator $\hat{\alpha}$ acting on V, in some continuous basis χ, by $[\hat{\alpha}]^{\chi}$ (using the same symbol as matrix representations). If $\hat{\alpha}$ acts on a vector $|\psi\rangle$ to give $|\psi\rangle'$, then its action in the χ space will be represented as

$$
\begin{aligned}
& & \psi'(\chi) &= [\hat{\alpha}]^{\chi}\,\psi(\chi) \\
\text{or} & & \langle\chi|\psi\rangle' &= [\hat{\alpha}]^{\chi}\,\langle\chi|\psi\rangle \\
\text{or} & & \langle\chi|\hat{\alpha}|\psi\rangle &= [\hat{\alpha}]^{\chi}\,\langle\chi|\psi\rangle
\end{aligned}
$$

Since the vector $|\psi\rangle$ in the above equation is arbitrary,

the role of the differential operator representation of an operator can be succinctly expressed using the following mnemonic:

$$
\langle\chi|\,\hat{\alpha} \;=\; [\hat{\alpha}]^{\chi}\,\langle\chi|
$$

After the next section, we will demonstrate what kind of quantization rules actually lead to such differential operator representations. Incidentally, in what follows we will often refer to a quantity like $\langle\chi\,|\hat{\alpha}|\,\chi'\rangle$ as a matrix element, borrowing the term from the finite, discrete basis case. Of course, for a basis labelled by a continuous index χ, strictly speaking, there really is no matrix to speak of.

In Table 7.1, we provide a *dictionary* that displays how one needs to carry out the translation as one moves from the discrete to the continuous basis description of QM.

---

[14]Loosely, a differential operator is a finite, linear combination of terms involving derivation operators and functions. A few common examples are

$$
\frac{d}{dx}, \quad \frac{\partial}{\partial x}, \quad x\frac{\partial}{\partial x}, \quad \sum_{i} x_i \frac{\partial}{\partial x_i}, \quad x^2 + \sin x, \quad 13
$$

Note that the differential operator can involve simple multiplicative functions like $x^2 + \sin x$ or even a constant like 13. In general, the differential operator acting on a space of functions of several variables will involve functions (of the different variables) and their partial derivation operators. A typical example is the operator $\sum_{i} x_i\,(\partial/\partial x_i)$ listed above.

TABLE 7.1: Discrete vs continuous spectra.

Discrete	Continuous
$\hat{\chi}$ has a discrete spectrum: $$\chi = \{\chi_1, \chi_2, \ldots, \chi_n\}$$	$\hat{\chi}$ has a continuous spectrum: $$\chi = [a, b]$$
Vectors in χ space are column matrices: $$[\psi]^\chi = (\langle\chi_i|\psi\rangle\,; i = 1, 2, \ldots, n)$$ $$[\psi]^\chi \in \mathbb{C}^n$$ $$\psi^i = \langle\chi_i|\psi\rangle$$	Vectors in χ space are complex functions: $$[\psi]^\chi = (\langle\chi|\psi\rangle\,; \chi \in [a, b])$$ $$[\psi]^\chi \in \mathbb{F}^c[a, b]$$ $$\psi(\chi) = \langle\chi|\psi\rangle$$
$\mathbb{C}^n$ forms a vector space under addition of complex column vectors and mutiplication of such vectors by a complex number	$\mathbb{F}^c[a, b]$ forms a vector space under addition of complex functions and mutiplication of such functions by a complex number
Vectors $[\chi_i]^\chi$ provide a standard basis: $$\psi^i = \sum_j \chi^i_j \psi^j$$ where $$\chi^i_j = \langle\chi_i|\chi_j\rangle = \delta^i_j$$	Functions $\chi(\chi')$ provide a standard basis: $$\psi(\chi') = \int_a^b d\chi \chi(\chi')\psi(\chi)$$ where $$\chi(\chi') = \langle\chi'|\chi\rangle = \delta(\chi - \chi')$$
Inner products are defined by $$\langle\phi|\psi\rangle = \sum_i \left(\phi^i\right)^* \left(\psi^i\right)$$ - always convergent for finite n The standard basis is orthonormal: $$\langle\chi_i|\chi_j\rangle = \delta^i_j$$	Inner products defined by $$\langle\phi|\psi\rangle = \int_a^b d\chi\,(\phi(\chi))^*\,(\psi(\chi))$$ - converges in $\mathbb{L}^2[a, b]$ The standard basis is orthonormal: $$\langle\chi'|\chi\rangle = \delta(\chi - \chi')$$
Abstraction proceeds as: $$|\psi\rangle = \sum_i |\chi_i\rangle\langle\chi_i|\psi\rangle$$ The column vectors $[\psi]^\chi$ are representations of $|\psi\rangle$ in the $\{|\chi_i\rangle\}$ basis	Abstraction proceeds as: $$|\psi\rangle = \int_a^b d\chi\,|\chi\rangle\langle\chi|\psi\rangle$$ The functions $\psi(\chi)$ are representations of $|\psi\rangle$ in the $\{|\chi\rangle\}$ basis
Representations of operators are square matrices on $\mathbb{C}^n$: $$\hat{\alpha}|\psi\rangle = |\psi\rangle'$$ is represented as $$[\hat{\alpha}]^\chi [\psi]^\chi = [\psi']^\chi$$ with $$[\hat{\alpha}]^\chi_{ij} = \langle\chi_i|\hat{\alpha}|\chi_j\rangle$$	Representations of operators are differential operators on $\mathbb{L}^2[a, b]$: $$\hat{\alpha}|\psi\rangle = |\psi\rangle'$$ is represented as $$[\hat{\alpha}]^\chi \psi(\chi) = \psi'(\chi)$$ where $$[\hat{\alpha}]^\chi : \langle\chi|\hat{\alpha} = [\hat{\alpha}]^\chi\langle\chi|$$

Modified Postulates

Two of our postulates will have to be reviewed and revised when observables with continuous spectra are involved. To see this, consider a Hermitian operator $\hat{\chi}$ having a continuous spectrum. Suppose that on a properly normalized state $|\psi\rangle$, an *imprecise measurement* of the observable $\hat{\chi}$ is made that measures whether the outcome χ belongs to some well defined interval $[\chi_0 - \epsilon, \chi_0 + \epsilon]$. From what we have learnt earlier about imprecise observables, we would expect the following rules to be obeyed:

1. **New state after a measurement**: If the outcome of the imprecise measurement is affirmative, then soon after the measurement the new state $|\psi ; \chi_0 \pm \epsilon\rangle$ will become

$$|\psi ; \chi_0 \pm \epsilon\rangle = \frac{\int_{\chi_0 - \epsilon}^{\chi_0 + \epsilon} d\chi \, |\chi\rangle \langle\chi|\psi\rangle}{\sqrt{\int_{\chi_0 - \epsilon}^{\chi_0 + \epsilon} d\chi \, |\langle\chi|\psi\rangle|^2}}$$

   This is essentially the *state collapse postulate* that was introduced earlier for the discrete case. It should be emphasized here that although the vectors $|\chi\rangle$ do not qualify as true quantum states (because they do not have the appropriate normalization), a measurement with inadequate resolution of $\hat{\chi}$ is still possible. In fact, the only measurements involving observables with continuous spectra are such imprecise measurements.

2. **Probability of outcomes**: The probability that the measurement of the imprecise observable yields an affirmative outcome is given by

$$P\left[|\psi\rangle \to |\psi ; \chi_0 \pm \epsilon\rangle\right] = \int_{\chi_0 - \epsilon}^{\chi_0 + \epsilon} d\chi \, |\langle\chi|\psi\rangle|^2$$

   This again is essentially the *Born rule* (for imprecise observables) that we have seen earlier for the discrete spectrum case. It should be noted that the inner product $\langle\chi|\psi\rangle$ has a slightly different interpretation in the present scenario.

*Here, $|\langle\chi|\psi\rangle|^2 \equiv |\psi(\chi)|^2$ describes the **probability density** instead of the probability; The probability $P\left[\psi \to [\chi, \chi + d\chi]\right]$ that an imprecise $\hat{\chi}$ measurement on a state $\psi(\chi)$ yields an outcome that ensures that χ lies in the interval $[\chi, \chi + d\chi]$ is given by*[15]

$$P\left[\psi \to [\chi, \chi + d\chi]\right] = |\psi(\chi)|^2 d\chi$$

---

[15]See the subsection on probability density function in appendix 'A'.

We notice a remarkable deviation from the discrete case: although the quantity $|\langle\chi|\psi\rangle|^2 = |\psi(\chi)|^2$ provides the probability for the event $\psi \to [\chi, \chi + d\chi]$, unlike the discrete case, the same quantity $|\psi(\chi)|^2 = |\langle\psi|\chi\rangle|^2$ cannot describe the probability for the reversed event $\chi \to [\chi; \psi]$. In fact, since the vectors $|\chi\rangle$ do not qualify as quantum states, the reversed event does not even exist.

The above rules are natural transcriptions of the corresponding postulates for the discrete case. However, since they cannot be deduced from them, *these rules have to go into the theory as independent postulates.*

We shall not list these rules in the "Postulates of QM" separately, but it is to be borne in mind that whenever observables with continuous spectra are involved, these will be the appropriate rules to use.

Quantization of Systems with Classical Analogues

Now we shall take up the issue of constructing quantum models of real life systems. It turns out that an overwhelmingly large class of quantum systems have *classical analogues*. By this we mean that for such systems we can frame a classical picture in our mind, so that it is possible to use the *vocabulary* of classical physics in the process of quantization. It is needless to mention that the classical model will not be the correct description of the system (which is why we seek a quantum description for it). It is not expected to lead to predictions that would match our experimental observations on the system. The framing of the classical picture is only an intermediate step to enable the quantum description to be formulated. The scheme essentially comprises laying down a consistent prescription to write down the quantization rules for a given classical system.

The essence of the preceding paragraph is that, when confronted with the task of formulating a quantum model for a real life system, one can try to[16]

1. *imagine an appropriate classical system, and*

2. *quantize the classical system according to the laid out prescription*[17].

If this process correctly reproduces all the experimental observations on the system, the quantization would be deemed to be correct. If it does not, one has to come up with some new quantization rule.

---

[16]Historically, this method evolved in the reverse route. Unsuccessful attempts of classical descriptions for certain systems lead people to tweak and twist the classical descriptions in order to explain the discrepancies of classical predictions with experimental observations. The *ad hoc* modifications which were implanted in the classical descriptions are now understood to be features of the quantized versions of those classical models.

[17]The prescription will be described shortly.

The language that we shall use for the classical description is that of a *classical Hamiltonian formalism*[18]. The corresponding quantization rule is called *canonical quantization*. In the Hamiltonian formalism, a system is described by two sets of basic variables: coordinates $\{q_i\}$ and momenta $\{p_i\}$. Each coordinate q_i necessarily comes with a conjugate momentum p_i, and i runs over all the coordinates. Every other observable (classically, one usually uses the term dynamical variable) is a function of the coordinates and momenta[19]. The time evolution of the system (which is essentially that of the coordinates and momenta) is governed by a function $H(q,p)$, called the Hamiltonian, through the equations

$$\frac{dq_i}{dt} = \frac{\partial H(q,p)}{\partial p_i}$$

$$\frac{dp_i}{dt} = -\frac{\partial H(q,p)}{\partial q_i}$$

where q and p stands collectively for all the coordinates and momenta. For all practical purposes (at least, for what we wish to discuss), one can take the Hamiltonian to be identical to the *energy* of the system, and we shall often use the terms Hamiltonian and energy interchangeably.

Now we turn to the quantization prescription of classical systems given in the Hamiltonian description. We start by laying it down for a particle whose motion is confined to one spatial dimension.

Particle living in one dimension

Classically, for this system there is one coordinate and one momentum. Let us denote them by x and p respectively. Every dynamical variable θ for the system will be written as a function $\theta(x, p)$. The quantization of this system involves the following assumptions and observations[20].

1. There exists an observable $\hat{x}$ called position corresponding to the classical coordinate x which has a spectrum $\{x : x \in (-\infty, +\infty)\}$. The observable $\hat{x}$ forms a CSCO by itself which acts on a vector space $\overline{V}$ being the linear span of the basis $\mathcal{B}^x = \{\,|x\rangle : x \in (-\infty, +\infty)\,\}$ where $|x\rangle$ are the eigenvectors of $\hat{x}$. Vectors $|\psi\rangle$ in $\overline{V}$ can be expanded formally as

$$|\psi\rangle = \int_{-\infty}^{+\infty} dx\,|x\rangle\,\langle x|\psi\rangle$$

---

[18] If you are not familiar with the formalism, do not worry, we do not assume any background.

[19] There can also be an explicit time dependence.

[20] Let us, for one last time, warn the reader that the treatment provided here is purely heuristic. The lack of rigor should be repaired when the necessary mathematical background has been acquired.

The x representations of vectors of this space are essentially functions of x. The standard jargon for this representation is "position space"[21]. So $\overline{V}$ is essentially the space of all functions of x. There exists a subspace V of $\overline{V}$ consisting of vectors whose x representations are square integrable functions. States of the quantum system belong to V. This space is endowed with an inner product

$$\langle \phi | \psi \rangle = \int_{-\infty}^{+\infty} dx\, \phi(x)^* \, \psi(x)$$

The basis vectors $|x\rangle$ are orthonormal in the extended sense:

$$\langle x' | x \rangle = \delta(x - x')$$

The representation of $\hat{x}$ follows trivially:

$$
\begin{aligned}
\langle x' | \hat{x} | x \rangle &= x\delta(x - x') \\
\langle x | \hat{x} &= x \langle x | \\
[\hat{x}]^x &= x
\end{aligned}
$$

that is, in its own space, the operator $\hat{x}$ can be represented as multiplicative operator[22].

2. There exists an observable $\hat{p}$ called momentum that corresponds to the classical momentum p. The representation of $\hat{p}$ in the x space is given by[23]

$$[\hat{p}]^x = -i\hbar \frac{d}{dx}$$

3. Every observable $\hat{\theta}$ of the *quantum* system corresponds to some dynamical variable $\theta(x, p)$ in the classical system. The operator corresponding to $\hat{\theta}$ is a function of the operators $\hat{x}$ and $\hat{p}$. The function $\hat{\theta}(\hat{x}, \hat{p})$ is given by

$$\hat{\theta}(\hat{x}, \hat{p}) = \theta(\hat{x}, \hat{p})$$

which is the same function by which the corresponding classical variable θ is connected to x and p.

---

[21] The phrase "coordinate space" is also used which often refers to a more general setting as the name suggests.

[22] You will have to prove this in the problem set provided at the end of this chapter. To see how this can be proved, check out the following subsection where the representation of the momentum operator is demonstrated to be a differential operator (described in the next item below). Note that a multiplicative operator is also, by definition, a trivial example of a differential operator.

[23] We shall demonstrate how this was arrived at shortly.

Sometimes, however, the construction may require minor adaptations. One such adaptation is **symmetrization**. This is required when products of classical variables corresponding to noncommuting operators need to be transcribed. For example, since $\hat{x}$ and $\hat{p}$ are noncommuting (which one can easily check), a term xp occurring in a dynamical variable is transcribed as $(1/2)(\hat{x}\hat{p} + \hat{p}\hat{x})$. Note that, in this example the symmetrization is necessary not merely to avoid unnecessary bias in the order of the operators $\hat{x}$ and $\hat{p}$ in the product, but for a much more important reason: the operator product $\hat{x}\hat{p}$ is not Hermitian! Unless we symmetrize, we do not have a quantum observable.

We mention, in particular, that the Hamiltonian operator $\hat{H}(\hat{x}, \hat{p})$, that governs the time evolution of the system, is given by

$$\hat{H}(\hat{x}, \hat{p}) \;=\; H(\hat{x}, \hat{p})$$

where $H(x, p)$ is the (duly symmetrized) classical Hamiltonian of the system.

Origin of the representation of the momentum operator

The most nontrivial and nonobvious component in the above prescription is perhaps the representation of the momentum operator in the position space[24]: $[\hat{p}]^x = -i\hbar(d/dx)$. Historically, there have been two routes to it. They are the famous *de Broglie hypothesis* and *canonical commutation relation*. Let us explicitly demonstrate how one derives $[\hat{p}]^x$ from each of them.

de Broglie Hypothesis The de Broglie hypothesis specifies the momentum eigenvectors in the position space as

$$\langle x|p\rangle \;=\; C\,e^{\frac{i}{\hbar}px}$$

where C is some constant of normalization and $\hbar$ is the Planck's constant[25] (which is there to make the theory fit observed data). It is then a trivial exercise to figure out the representation of the momentum operator. By mere inspection it is evident that a x space differential operator that has $C\exp\{(i/\hbar)px\}$ as its eigenvector with p as the corresponding eigenvalue is $-i\hbar(d/dx)$.

---

[24]The representation of any operator in a space other than the one constructed out of its own eigenvectors is *always* nontrivial.

[25]It is only after several re-interpretations that the meaning of de Broglie's hypothesis became clear. The original hypothesis made the assertion that there should be waves associated with particles whose momenta p are related to the wave lengths λ of the waves by $p = h/\lambda$. Historically, people called this the *"wave particle duality"*. Note that, since the Hamiltonian for the free particle is $\hat{H} = \hat{p}^2/2m$ (where m is the mass), the momentum eigenstates are stationary, and a momentum eigenstate after time t is $\langle x|p(t)\rangle = C\exp\{(i/\hbar)(px - Et)\}$ which is indeed a progressive wave!

The fact that the eigenvectors of the momentum operator $\hat{p}$ are given in the position space by *hypothesis,* means that here the solution to the eigenvalue problem of the momentum operator $\hat{p}$ is *assumed,* and the representation is *deduced!* The point is, once the representations of the position and momentum operators are available, the representations of all other observables can be constructed, and one can proceed with the usual program of solving their eigenvalue problems to find the answers to the target questions of QM.

Canonical Commutation Relation[26] In this prescription one hypothetically specifies the commutation relation between $\hat{x}$ and $\hat{p}$:

$$[\hat{x}, \hat{p}] = i\hbar$$

From here, one can derive the representation of either of the operators in the space of its conjugate. We show how the x representation of $\hat{p}$ can be derived.

From the above commutation relation, we have

$$\langle x' |[\hat{x}, \hat{p}]| x \rangle = \langle x' |i\hbar| x \rangle$$

The matrix element of the commutator on the left hand side is

$$\langle x' |[\hat{x}, \hat{p}]| x \rangle = (x' - x) \langle x' |\hat{p}| x \rangle$$

But the right hand side gives

$$\langle x' |i\hbar| x \rangle = i\hbar\delta (x - x')$$

Hence by equating the two sides, we have

$$\langle x' |\hat{p}| x \rangle = i\hbar\frac{\delta (x - x')}{(x' - x)}$$

$$= i\hbar\frac{d}{dx}\delta (x - x')$$

where we have used a delta function identity:

$$\frac{\partial}{\partial x}\delta (x) = -\frac{\delta (x)}{x}$$

Now, we can write for an arbitrary state $|\psi\rangle$

$$\langle x' |\hat{p}| \psi \rangle = \int_{-\infty}^{+\infty} dx \, \langle x' |\hat{p}| x \rangle \langle x|\psi\rangle$$

$$= \int_{-\infty}^{+\infty} dx \left[i\hbar\frac{d}{dx}\delta (x - x') \right] \psi (x)$$

$$= -i\hbar\frac{d}{dx'}\psi (x')$$

i.e., $$\langle x |\hat{p}| \psi \rangle = -i\hbar\frac{d}{dx} \langle x|\psi\rangle$$

---

[26]This hypothesis was inspired by the so called *Poisson bracket* formulation of classical mechanics.

Here, in the third line, we have used another delta function identity:

$$\int dx \left[\frac{d}{dx}\delta\left(x\right) \right] \psi\left(x\right) = -\frac{d}{dx}\psi\left(x\right)\Big|_{x=0}$$

(where the range of the integral is assumed to include the point $x = 0$), and in the last line we have switched the variable from x' to x. Since the derived expression for $\langle x\,|\hat{p}|\,\psi\rangle$ is true for all states $|\psi\rangle$, it implies

$$\langle x|\,\hat{p} = -i\hbar\frac{d}{dx}\langle x|$$

$$[\hat{p}]^x = -i\hbar\frac{d}{dx}$$

Justification for $\hat{\theta}\left(\hat{x},\,\hat{p}\right) = \theta\left(\hat{x},\,\hat{p}\right)$

Although the rule: $\hat{\theta}\left(\hat{x},\,\hat{p}\right) = \theta\left(\hat{x},\,\hat{p}\right)$ does not look very strange and far fetched, an inquisitive mind is bound to wonder why it works. Classical physics should not tell us what the quantum theory should be like. We expect classical physics to *emerge* from quantum theory and not the other way around[27]. Be that as it may, a more reasonable and legitimate question to ask is probably, what are the facts from which this rule could be inferred? A strong evidence for the quantization rule is provided by the fact that expectation values of various observables resemble classical relationships for *arbitrary* states. For example, when $V\left(x\right)$ is the potential energy, if the relationship

$$\langle\psi|\,\hat{H}\,|\psi\rangle = \langle\psi|\,\frac{\hat{p}^2}{2m}\,|\psi\rangle + \langle\psi|\,V\left(\hat{x}\right)|\psi\rangle$$

between the Hamiltonian, momentum and position expectations hold for any arbitrary state $|\psi\rangle$, it would strongly suggest that the relationship

$$\hat{H} = \frac{\hat{p}^2}{2m} + V\left(\hat{x}\right)$$

is true at the operator level.

Position and momentum states

We have seen that eigenvectors $|x\rangle$ of the position operator $\hat{x}$ are not normalized to unity (actually $\langle x|x\rangle$ is infinite). So the vectors $|x\rangle$ cannot qualify as quantum states. This means that *a quantum particle with definite position does not exist!* The vectors $|x\rangle$ serve only to provide an orthonormal basis (with orthonormality in the extended sense) of an inner product space which

---

[27]It might be worth considering whether this quantization rule is actually a manifestation of the fact that the classical physics is actually getting cooked up by the underlying quantum machinery. But nobody has proved it as yet.

contains the properly normalizable vectors $|\psi\rangle$. It is with this understanding that we write the equations

$$|\psi\rangle = \int_{-\infty}^{+\infty} dx \, |x\rangle \langle x|\psi\rangle$$

$$\langle x'|x\rangle = \delta(x - x')$$

(where $\langle x|\psi\rangle$ is square integrable). It is easy to see that the eigenvectors $|p\rangle$ of the momentum operator $\hat{p}$ are also not normalizable to unity

$$\langle p|p\rangle = \int_{-\infty}^{+\infty} dx \, \left| C e^{\frac{i}{\hbar}px} \right|^2 = |C|^2 \int_{-\infty}^{+\infty} dx \longrightarrow \infty$$

However, just as the position eigenvectors, they can also serve to provide an orthonormal basis of the Hilbert space where states live. This follows from Fourier's theorem: A *reasonably well behaved* function $\psi(x)$ can be expressed as the following integral[28]:

$$\psi(x) = \frac{1}{\sqrt{2\pi\hbar}} \int_{-\infty}^{+\infty} dp \, \tilde{\psi}(p) \, e^{\frac{i}{\hbar}px}$$

where $\tilde{\psi}(p)$ is called the Fourier transform of $\psi(x)$. In a representation free language, we can write this as

$$|\psi\rangle = \int_{-\infty}^{+\infty} dp \, |p\rangle \langle p|\psi\rangle$$

where $\langle p|\psi\rangle = \tilde{\psi}(p)$. It is easy to see that the vectors $|p\rangle$ respect orthonormality in the extended sense as well. Using the expression for $\langle x|p\rangle$, we have

$$\langle p'|p\rangle = \frac{1}{2\pi\hbar} \int_{-\infty}^{+\infty} dx \, e^{\frac{i}{\hbar}(p-p')x}$$

The right hand side is a well known representation of the Dirac delta function[29]. Thus

$$\langle p'|p\rangle = \delta(p - p')$$

Schrödinger equation in position space

The representations of all observables can be easily constructed in terms of the representations of $\hat{x}$ and $\hat{p}$. Let us illustrate this for the Hamiltonian of our system:

$$\hat{H} = \frac{1}{2m}\hat{p}^2 + V(\hat{x})$$

---

[28] By "reasonably well behaved" we mean, functions that are well behaved enough to admit Fourier transform. Please check what comprises a set of sufficient conditions for Fourier transforms to exist; see, for example, (Boas 1980).

[29] This can be found in any reference on the Dirac delta function; see, for example, (Cohen-Tannoudji et al. 1977, Appendix II).

where $V(x)$ is assumed to be a potential under which the classical particle moves. The coordinate space representation of the momentum operator $\hat{p}$ is given by

$$\langle x|\,\hat{p} = -i\hbar \frac{d}{dx}\langle x|$$

$$\langle x|\,\hat{p}\,|\psi\rangle = -i\hbar \frac{d}{dx}\langle x|\psi\rangle$$

where $|\psi\rangle$ is an arbitrary state. Now let us try to figure out what the position space representation of the operator $\hat{p}^2$ would be. We have

$$\langle x|\,\hat{p}^2\,|\psi\rangle = \langle x|\,\hat{p}\hat{p}\,|\psi\rangle$$

$$= -i\hbar \frac{d}{dx}\langle x|\,\hat{p}\,|\psi\rangle$$

$$= -\hbar^2 \frac{d^2}{dx^2}\langle x|\psi\rangle$$

Since this is true for an arbitrary state $|\psi\rangle$, we can write

$$\langle x|\,\hat{p}^2 = -\hbar^2 \frac{d^2}{dx^2}\langle x|$$

Along the same lines we can show that[30]

$$\langle x|\,\hat{x} = x\,\langle x| \quad\Longrightarrow\quad \langle x|\,V(\hat{x}) = V(x)\,\langle x|$$

The representation of the Hamiltonian will then be given by

$$\langle x|\,\hat{H} = \left(-\frac{\hbar^2}{2m}\frac{d^2}{dx^2} + V(x)\right)\langle x|$$

Now, the time evolution of the system is governed by the Schrödinger equation:

$$i\hbar \frac{d\,|\psi(t)\rangle}{dt} = \hat{H}\,|\psi(t)\rangle$$

which for our system becomes

$$i\hbar \frac{d\,|\psi(t)\rangle}{dt} = \left(\frac{1}{2m}\hat{p}^2 + V(\hat{x})\right)|\psi(t)\rangle$$

If we cast this equation in position space, we have

$$i\hbar \frac{d\,\langle x|\psi(t)\rangle}{dt} = \langle x|\,\frac{1}{2m}\hat{p}^2\,|\psi(t)\rangle + \langle x|\,V(\hat{x})\,|\psi(t)\rangle$$

---

[30]To see this, imagine that $V(x)$ is a polynomial in x.

Writing $\langle x|\psi(t)\rangle = \psi(x,t)$, and using the x representations of the operators $\hat{p}^2$ and $V(\hat{x})$, we get the Schrödinger equation in position space for a particle moving in one dimension:

$$i\hbar\frac{\partial\psi(x,t)}{\partial t} = \left(-\frac{\hbar^2}{2m}\frac{\partial^2}{\partial x^2} + V(x)\right)\psi(x,t)$$

Since the representation of the state vector in the position space is a function of x, it is clear that the Schrödinger equation is now a partial differential equation in x and t.

Solution of the Schrödinger equation in a simple case: particle in a 1-dimensional box

Although, the target questions of QM continue to be eigenvalue problems, the technology for solving the eigenvalue problems in the continuum are completely different. Since the representations of observables are now differential operators, the eigenvalue equations show up as differential equations. To illustrate this, in this section, we shall obtain the solution of the Schrödinger equation for a simple system. The classical analogue of the system that we wish to describe is a particle confined to a *one dimensional box*. Such a particle can be assumed to be moving under the potential

$$V(x) = \begin{cases} 0 & 0 \le x \le a \\ \infty & \text{otherwise} \end{cases}$$

where x is, as always, the Cartesian coordinate denoting position, and 'a' is the extent of the box. This is a conservative system and, therefore, as we have explained in the last chapter, the solution will essentially be reduced to solving the eigenvalue problem of the Hamiltonian.

Assuming the particle to be of mass m, the Hamiltonian of this particle will be given by

$$\hat{H} = \frac{1}{2m}\hat{p}^2 + V(\hat{x})$$

The representation of the Hamiltonian in position space will be expressed as

$$\langle x|\frac{1}{2m}\hat{p}^2 + V(\hat{x})|\psi\rangle = \left(-\frac{\hbar^2}{2m}\frac{d^2}{dx^2} + V(x)\right)\langle x|\psi\rangle$$

The eigenvalue equation of the Hamiltonian

$$\hat{H}|E\rangle = E|E\rangle$$

when cast in position space, gives us the differential equation

$$\langle x|\frac{1}{2m}\hat{p}^2 + V(\hat{x})|E\rangle = E\langle x|E\rangle$$

i.e., $$\left(-\frac{\hbar^2}{2m}\frac{d^2}{dx^2} + V(x)\right)E(x) = EE(x)$$

which has a form of the familiar differential equation

$$\left[\frac{d^2}{dx^2} + \kappa^2 \right] E_\kappa\left(x\right) = 0$$

$$\text{with} \quad \kappa^2 = \frac{2mE}{\hbar^2}$$

in the range $0 \leq x \leq a$. The general solution is well known:

$$E_\kappa\left(x\right) = A \sin \kappa x + B \cos \kappa x$$

where A and B are the constants of integration. Outside the box ($0 > x > a$), the wave function must be zero owing to the infinite potential there. Now, it so happens that the wave functions (i.e., the solution to our differential equation) must be continuous everywhere[31]. Imposing this condition at the boundary $x = 0$ gives $B = 0$, and at $x = a$ gives

$$\kappa a = n\pi$$

with[32] $n = \pm 1, \pm 2, \ldots$. This gives the allowed values of the energy[33]:

$$E_n = \frac{\pi^2 \hbar^2 n^2}{2ma^2}$$

where we have now chosen to write the energy with a subscript n as E_n. Thus the *eigenfunctions* read[34]

$$E_n\left(x\right) = A \sin\left(\frac{n\pi}{a}x\right)$$

The constant A is determined by the normalization condition

$$\int_0^a |E_n\left(x\right)|^2 = 1$$

which gives $A = \sqrt{2/a}$. Hence, finally, the normalized energy eigenfunctions are

$$E_n\left(x\right) = \sqrt{\frac{2}{a}} \sin\left(\frac{n\pi}{a}x\right)$$

---

[31] This can actually be proven rigorously from the nature of the differential equation. In the problem set that follows, you will be asked to convince yourself of this fact using a simple analysis (although, in a nonrigorous way).

[32] The possibility $n = 0$ is excluded for the same reason $A = 0$ is excluded. Can you see why? Yes, the wave function in these cases will be zero everywhere. But so what?

[33] Note that the eigenvalues are determined from the boundary conditions here. As you will see in the next chapter, the eigenvalues will often be obtained from other constraints that the eigenvectors will be expected to obey. In the continuum, they will not be obtainable through a general prescription as was possible for the discrete, finite dimensional case where they were provided by the solution to the characteristic equation.

[34] In the continuum, *eigenvectors* are often referred to as *eigenfunctions*.

These wave functions constitute an orthonormal basis for the state space of our system.

Now, we have learnt in the last chapter how to write a formal solution to the Schrödinger equation (i.e., the quantum mechanical state at an arbitrary time) $|\psi(t)\rangle$ in terms of the energy eigenvalues and eigenfunctions. Using this result, we write

$$|\psi(t)\rangle = \sum_n |E_n\rangle \langle E_n|\psi(0)\rangle \, e^{-\frac{i}{\hbar}E_n t}$$

In the position space, this takes the form

$$\psi(x,t) = \sum_n E_n(x) \, c_n e^{-\frac{i}{\hbar}E_n t}$$

where $\qquad c_n = \langle E_n|\psi(0)\rangle$

So we see that once the energy eigenvalue problem has been solved, one merely needs to compute the expansion coefficients c_n of the initial state $\psi(x,0)$ in the energy basis to write down the full solution. The orthonormality of the basis vectors $E_n(x)$ for our system gives

$$c_n = \int_0^a dx \, E_n(x) \, \psi(x,0)$$

noting that the $E_n(x)$ are real[35].

Particles living in higher dimensions

A classical particle moving in two dimensions can be described by two Cartesian coordinates x, y and their corresponding momenta p_x, p_y. This system can be quantized by specifying a CSCO comprising two position observables $\hat{x}$ and $\hat{y}$ (instead of just $\hat{x}$ that constituted a CSCO by itself in one dimension) both having continuous spectra. The operators act on a vector space $\overline{V}$ spanned by the basis $\mathcal{B}^{xy} = \{ |x,y\rangle : x,y \in (-\infty, +\infty) \}$ where $|x,y\rangle$ are common eigenvectors of $\hat{x}$ and $\hat{y}$. Vectors in this space can be expanded formally in this basis as

$$|\psi\rangle = \int_{-\infty}^{+\infty} \int_{-\infty}^{+\infty} dx dy \, |x,y\rangle \langle x,y|\psi\rangle$$

The position representation of $|\psi\rangle$ are functions of x and y: $\langle x,y|\psi\rangle = \psi(x,y)$. Quantum states live in the subspace V of $\overline{V}$ whose position representation comprises square integrable functions of x and y and the space V is endowed with an inner product in the usual way:

$$\langle \phi|\psi\rangle = \int_{-\infty}^{+\infty} \int_{-\infty}^{+\infty} dx dy \, \phi(x,y)^* \, \psi(x,y)$$

---

[35] This standard technique, of using an inner product to pick out the expansion coefficients in an orthonormal basis expansion is known as *Fourier's trick* in the continuum.

The orthonormality condition on the basis $\mathcal{B}^{xy}$ reads

$$\langle x', y' | x, y \rangle = \delta \left(x - x' \right) \delta \left(y - y' \right)$$

In the (x, y) representation, the $\hat{x}$ and $\hat{y}$ observables become multiplicative operators:

$$[\hat{x}]^{x,y} = x \qquad \text{and} \qquad [\hat{y}]^{x,y} = y$$

The representations of the momenta in the (x, y) space are given by

$$[\hat{p}_x]^{x,y} = -i\hbar \frac{\partial}{\partial x} \qquad \text{and} \qquad [\hat{p}_y]^{x,y} = -i\hbar \frac{\partial}{\partial y}$$

Every observable $\hat{\theta}$ corresponds to some classical dynamical variable $\theta \left(x, p_x, y, p_y \right)$, and is given by

$$\hat{\theta} = \theta \left(\hat{x}, \hat{p}_x, \hat{y}, \hat{p}_y \right)$$

Multiparticle systems

Multiparticle systems can be quantized in exactly the same way as a particle living in higher dimensions. For example, two classical particles moving in one dimension can be described by a prescription that can be obtained simply by replacing x, y, p_x and p_y above by x^1, x^2, p^1 and p^2 respectively, where the numbers '1' and '2' in the superscripts are used to label the two particles.

The extension of this recipe to more general scenarios (more particles and dimensions) is obvious[36].

Problems

1. Consider a particle of mass m moving in one dimension under a potential $V(x)$.

   (a) Show that the representation of the position operator in position space is given by[37]

   $$[\hat{x}]^x \psi(x) = x\psi(x)$$

---

[36] It is also not hard to imagine application of this prescription to other classical systems such as classical fields. This has actually been done and has led to whole new subject called Quantum Field Theory, which provides the basic framework for describing elementary particles (the fundamental building blocks that everything is made up of). But that is a story for another day.

[37] By the way, if this looks like an eigenvalue equation to you, you should look again and convince yourself that it certainly is not!

(b) Establish the Hermiticity of the momentum operator $[\hat{p}]^x = -i\hbar\,(d/dx)$ from the definition of Hermiticity in the x space: an operator $\hat{O}$ is Hermitian if its x representation $\left[\hat{O}\right]^x$ obeys the condition

$$\int_{-\infty}^{+\infty} dx\, f(x)^* \left[\hat{O}\right]^x g(x) = \int_{-\infty}^{+\infty} dx\, \left(\left[\hat{O}\right]^x f(x)\right)^* g(x)$$

where $f(x)$ and $g(x)$ are any two functions in the domain. Assume that $f(x)$ and $g(x)$ are localized (i.e., they vanish as $x \to \pm\infty$).

(c) Using separation of variables, show that a solution to the Schrödinger equation in the position space is

$$\psi_n(x,t) = e^{-\frac{i}{\hbar}E_n t} U_n(x)$$

where $U_n(x)$ is an eigenfunction of the Hamiltonian operator with eigenvalue E_n. These solutions are called *separable solutions*. Argue that every solution $\psi(x,t)$ of the Schrödinger equation is a linear superposition of the separable solutions:

$$\psi(x,t) = \sum_n c_n \psi_n(x,t)$$

where the coefficients c_n are given by

$$c_n = \int_{-\infty}^{+\infty} dx\, U_n^*(x)\, \psi(x,0)$$

(d) From the generalized uncertainty principle, show that

$$\sigma_x \sigma_p \geq \frac{\hbar}{2}$$

where σ_x and σ_p are the standard deviations in position and momentum respectively.

(e) Starting from the canonical commutation relation, prove the following commutation relations:

$$[\hat{x}, \hat{p}^n] = i\hbar n \hat{p}^{n-1} \qquad \text{and} \qquad [\hat{p}, \hat{x}^n] = -i\hbar n \hat{x}^{n-1}$$

(f) Show that

    i. the time evolution of the expectation values of the position and momentum observables, $\hat{x}$ and $\hat{p}$, obey the relations

$$\frac{d\langle\hat{x}\rangle}{dt} = \frac{\langle\hat{p}\rangle}{m} \qquad \text{and} \qquad \frac{d\langle\hat{p}\rangle}{dt} = -\frac{d\langle V(\hat{x})\rangle}{dx}$$

    Incidentally, these relationships, which demonstrate that the expectation values obey classical relationships, are known as *Ehrenfest's theorems*.

ii. in the Heisenberg picture, the time evolution of the position
and momentum observables, $\hat{x}_H$ and $\hat{p}_H$, obey the relations

$$\frac{d\hat{x}_H}{dt} = \frac{\hat{p}_H}{m} \quad \text{and} \quad \frac{d\hat{p}_H}{dt} = -\frac{dV(\hat{x}_H)}{dx}$$

(g) Obtain an expression for the differential rate of change of the probability dP/dt of finding the particle in some finite interval $[a, b]$. Interpret your result in terms of a probability flow.

(Hint: Start from the Schrödinger equation and its complex conjugate equation. What you will see on the right hand side is the divergence of a function. How will you call it?)

(h) Show that the wave function is always continuous and its first derivative (with respect to position) is also continuous if the potential has, at worst, finite discontinuities[38]. What will be the continuity condition if the potential is of the form

$$V(x) = -C\,\delta(x - x_0)$$

(i) Investigate how the results of this problem may be generalized for higher dimensions and multi-particle systems.

2. A particle of mass m is confined in a 1-dimensional box of size a.

(a) Check that the (normalized) energy eigenstates constitute an orthonormal basis for its state space.

(b) Suppose the particle is in an initial state

$$|\psi(0)\rangle = A\,[1\,|E_1\rangle + 2\,|E_2\rangle + 3\,|E_3\rangle]$$

where A is an appropriate constant that normalizes[39] $|\psi(0)\rangle$.

i. Find out the normalization constant A

ii. Determine the state $|\psi(t)\rangle$ at time t.

iii. What is the probability that at time t, in a measurement of position, the particle will be found in the position interval $[0, a/4]$? What would be your answer if immediately prior to the position measurement, an energy measurement is performed that yields an outcome E_2?

iv. Suppose an imprecise measurement is designed to tell us whether the particle is in the left half of the box $x \in [0, a/2]$ or the right half $x \in (a/2, a]$. If, at time t, the outcome of this measurement indicates that the particle is in the right half of the box, write down the state of the particle soon after the measurement.

[38] These are known as *wave function continuity conditions*.
[39] You might want to try using a computer algebra system for some parts of this problem.

(c) Assume that the particle starts out from an initial state that is given by the position space wave function

$$\psi(x,0) = C\,x\,(a - x)$$

    i. Find out the normalization constant C.

    ii. Suppose an imprecise measurement of energy is made on the system that detects whether the energy is even or odd (i.e., whether the quantum number n occurring in the expression for the energy is even or odd). Deduce the probability of the two possible outcomes.

    iii. If an imprecise measurement of energy made at time t yields that the energy $E < 16\pi^2\hbar^2/2m\,a^2$, what is the final state soon after the measurement?

3. Consider a particle of mass m confined to a rectangular, 2-dimensional box of sides a and b (along x and y axes respectively of a Cartesian system).

    (a) Write down the Hamiltonian for this system in position (x, y) space.

    (b) Write down the spectrum and normalized eigenstates of this Hamiltonian using the results of the previous problem.

    (c) Your Hamiltonian $\hat{H}$ should consist of an x dependent part, say $\hat{H}_x$, and a y dependent part, say $\hat{H}_y$, so that $\hat{H} = \hat{H}_x + \hat{H}_y$. Show that the pair $\left\{\hat{H}_x, \hat{H}_y\right\}$ forms a CSCO. Will $\hat{H}$ form a CSCO? Study the degeneracy in the case of a square box: $a = b$.

    (d) Assume that the system starts from an initial state

$$\psi(x, y; 0) = C\,x\,(a - x)\cos\frac{2\pi y}{b}\sin\frac{3\pi y}{b}$$

    where C is the normalization constant.

    i. Find out the state $\psi(x, y; t)$ at time t.

    ii. What are the possible measurement outcomes of $\hat{H}_y$ and the probabilities of the different possibilities?

4. A particle is moving in a closed circular loop having a radius a. The position of the particle can be described by using the polar angle ϕ. In the quantized version, the observable $\hat{\phi}$ forms a CSCO by itself having a spectrum $\{\phi \in [0, 2\pi]\}$, and $\{|\phi\rangle \; ; \; \phi \in [0, 2\pi]\}$ is the corresponding orthonormal basis that spans the state space. Now let us define an observable $\hat{G}$ by its action in the ϕ space as

$$\langle\phi|\,\hat{G} = i\hbar\frac{d}{d\phi}\,\langle\phi|$$

(a) Prove that $\hat{G}$ is Hermitian in the domain $\phi \in [0, 2\pi]$.

(b) Determine the spectrum of $\hat{G}$ and the corresponding orthonormal eigenstates.

(Hint: The eigenstates $G(\phi)$ must obey periodic boundary conditions: $G(\phi) = G(\phi + 2\pi)$ on physical grounds.)

(c) Does $\hat{G}$ form a CSCO?

(d) Can you guess a classical dynamical variable that $\hat{G}$ corresponds to?

(e) Construct the Hamiltonian $\hat{H}$ for the system and express it in terms of $\hat{G}$.

(f) Will $\hat{H}$ form a CSCO?

(g) Can you normalize the eigenfunctions if the loop is an ellipse with semi-major axis a and semi-minor axis[40] b?

5. The parity operator $\hat{\Pi}$ is defined by its action on a basis comprising the position vectors $|\vec{r}\rangle$ (where $|\vec{r}\rangle \equiv |x, y, z\rangle$) as

$$\hat{\Pi} |\vec{r}\rangle = |-\vec{r}\rangle$$

Show that

(a) the position space representation of $\hat{\Pi}$ can be described as

$$\langle \vec{r} | \hat{\Pi} = \langle -\vec{r} |$$

(b) $\hat{\Pi}$ is both Hermitian and unitary.

(c) the eigenvalues of $\hat{\Pi}$ are $+1$ or -1. The corresponding eigenvectors are called *even* and *odd* respectively.

(d) the eigenspaces associated with the eigenvalues of $\hat{\Pi}$ are orthogonal and the vector space on which $\hat{\Pi}$ is defined is a direct sum of the even and odd eigenspaces[41].

---

[40]Strangely, it turns out that it is impossible to do this exactly. Look up the literature to give an approximate expression.

[41]Two subspaces are orthogonal if every vector of one subspace is orthogonal to every vector of the other. Although these propositions can be proved directly, another instructive way to show them is to consider the operators

$$\hat{E} = \frac{1}{2}\left(\hat{I} + \hat{\Pi}\right) \qquad \hat{O} = \frac{1}{2}\left(\hat{I} - \hat{\Pi}\right)$$

and show that the operators $\hat{E}$ and $\hat{O}$ are projectors onto the even and odd eigenspaces. It is then a simple algebraic exercise to show that

$$\hat{E}\hat{O} = \hat{O}\hat{E} = \hat{0} \qquad \text{and} \qquad E + \hat{O} = \hat{I}$$

where $\hat{0}$ and $\hat{I}$ is the null operator and identity operator respectively. You may look up the section on *direct sum* and *projection operators* in appendix 'B' if required.

(e) if an observable $\hat{A}$ is *even* (i.e., it remains invariant under parity[42]):

$$\hat{\Pi}\hat{A}\hat{\Pi} = \hat{A}$$

it has a complete set of eigenvectors of definite parity[43]. What can you say about the degeneracy of the eigenvalue of an even operator $\hat{A}$ if its eigenvector does not have a definite parity?

(f) if an observable $\hat{A}$ is *odd:*

$$\hat{\Pi}\hat{A}\hat{\Pi} = -\hat{A}$$

its matrix elements between states which have a definite parity vanish. How would your result be modified if the operator was even?

---

[42] You will have already noticed that the parity operator is its own inverse.

[43] Check that the eigenstates of the Hamiltonian of the particle in a box (problem 2) have definite parity. Can you infer this without an explicit check?

Chapter 8

Three Archetypal Eigenvalue Problems

In the previous chapter, we have introduced the basic scheme of things for the description of quantum systems that involve inner product spaces whose members can be expressed using bases that form a continuum. In this chapter we shall discuss three, very well known, quantum mechanical eigenvalue problems for systems that belong to this class: i) energy of a 1-dimensional harmonic oscillator, ii) angular momentum and iii) internal energy of the hydrogen atom[1]. These problems are treated in almost all books on QM. There are many reasons why these examples have become so important. Among other things, they are among the few real life problems in QM that can actually be solved exactly. For most problems that one encounters in nature, one has to resort to sophisticated approximation techniques[2]. The problems that we discuss here will introduce some of the basic methods used in solving typical, real life, quantum mechanical eigenvalue problems.

Harmonic Oscillator

The classical 1-dimensional harmonic oscillator is a system comprising a particle of mass m moving in one dimension along some axis, say x-axis, under a harmonic potential $V(x) = (1/2) m\omega^2 x^2$ where ω is the natural frequency[3]. The *quantum* harmonic oscillator is a quantum system that is quantized using an intermediate *classical* harmonic oscillator.

The Hamiltonian operator for the quantum system that controls its time evolution is given by

$$\hat{H} = \frac{1}{2m}\hat{p}^2 + \frac{1}{2}m\omega^2 \hat{x}^2$$

---

[1] More specifically, we are referring to *orbital* angular momentum here, although, we shall also discuss, *generalized* angular momentum using an algebraic method which is not tied to continuous bases.

[2] Unfortunately we will not discuss these techniques in this book.

[3] For a detailed account of this very important system, see (Cohen-Tannoudji et al. 1977). See, in particular, *complement A* of chapter five where several applications of this model to actual physical systems have been discussed.

Since the Hamiltonian has no explicit time dependence, the time evolution of the system reduces to the solution of the eigenvalue problem of the Hamiltonian, as we have explained before. In this section, we shall solve this eigenvalue problem to determine the spectrum and the corresponding wave functions.

Analytical method

The coordinate space representation of the Hamiltonian follows from the representations of $\hat{x}$ and $\hat{p}$ in coordinate space:

$$\langle x| \frac{1}{2m}\hat{p}^2 + \frac{1}{2}m\omega^2\hat{x}^2 |\psi\rangle = \left(-\frac{\hbar^2}{2m}\frac{d^2}{dx^2} + \frac{1}{2}m\omega^2 x^2 \right) \langle x|\psi\rangle$$

Now, the eigenvalue equation for $\hat{H}$ is

$$\hat{H}|E\rangle = E|E\rangle$$

Expressing this equation in the x representation we have

$$\langle x| \frac{1}{2m}\hat{p}^2 + \frac{1}{2}m\omega^2\hat{x}^2 |E\rangle = E\langle x|E\rangle$$
$$\left(-\frac{\hbar^2}{2m}\frac{d^2}{dx^2} + \frac{1}{2}m\omega^2 x^2 \right) \langle x|E\rangle = E\langle x|E\rangle$$

The above equation can be identified with a *standard differential equation*:

$$-\frac{d^2 y_n}{d\rho^2} + \rho^2 y_n = (2n+1)\, y_n$$

if we write $y_n(\rho) = \langle x|E\rangle$ with the simple substitutions

$$x = \rho\sqrt{\frac{\hbar}{m\omega}} \quad \text{and} \quad E = \left(n + \frac{1}{2} \right)\hbar\omega$$

This equation is usually discussed in connection with Hermite's differential equation; see, for example, (Margenau and Murphy 1956). It turns out that, for the solutions of this differential equation to be square integrable, n must be a nonnegative integer. This gives the eigenvalues E_n of $\hat{H}$:

$$E_n = \left(n + \frac{1}{2} \right)\hbar\omega \quad \text{with} \quad n = 0,1,2,3,\ldots$$

Writing $y_n(\rho) = \langle \rho|n\rangle$, the corresponding normalized eigenvectors of $\hat{H}$ are

$$\langle \rho|n\rangle = \sqrt{\frac{1}{x_0\sqrt{\pi}\, 2^n\, n!}}\; H_n(\rho)\, e^{-\frac{1}{2}\rho^2}$$

where $x_0 = \sqrt{\hbar/m\omega}$ and $H_n(\rho)$ are n-th order polynomials known as *Hermite polynomials* in the mathematical literature[4]. These are well tabulated in most handbooks of mathematical functions. The **Hermite polynomials** may be defined by the, so called, Rodrigues' formula:

$$H_n(\rho) = (-1)^n e^{x^2} \left(\frac{d}{d\rho}\right)^n e^{-x^2}$$

Algebraic method

It turns out that the eigenvalue spectra can sometimes be obtained by using purely algebraic means. The method also provides a way to obtain the eigenvectors by solving a much simpler equation than the full eigenvalue equation. We demonstrate this method below.

Eigenvalue spectrum

Let us define operators $\hat{a}$ and $\hat{a}^\dagger$ by

$$\hat{a} = \frac{1}{\sqrt{2}} \left(\sqrt{\frac{m\omega}{\hbar}}\hat{x} + \frac{i}{\sqrt{m\hbar\omega}}\hat{p}\right)$$

$$\hat{a}^\dagger = \frac{1}{\sqrt{2}} \left(\sqrt{\frac{m\omega}{\hbar}}\hat{x} - \frac{i}{\sqrt{m\hbar\omega}}\hat{p}\right)$$

and another operator $\hat{N}$ by

$$\hat{N} = \hat{a}^\dagger \hat{a}$$

The operator $\hat{N}$ is called **number operator,** and the operators $\hat{a}$ and $\hat{a}^\dagger$ are called **ladder operators,** for reasons that will become apparent shortly. Using these operators we can rewrite the Hamiltonian as

$$\hat{H} = \left(\hat{N} + \frac{1}{2}\right)\hbar\omega$$

Now, the canonical commutation relation between $\hat{x}$ and $\hat{p}$ is given by

$$[\hat{x}, \hat{p}] = i\hbar$$

It is easy to check that

$$[\hat{x}, \hat{p}] = i\hbar \quad \Longrightarrow \quad [\hat{a}, \hat{a}^\dagger] = 1$$

We shall prove that the eigenvalues of $\hat{N}$ are nonnegative integers:

$$\hat{N}|n\rangle = n|n\rangle \quad \text{with} \quad n = 0, 1, 2, \ldots$$

---

[4]A detailed solution of this equation (by making use of the *power series* method) is worked out in many standard undergraduate text books. See, for example, (Griffiths 2005).

The eigenvalues E_n of $\hat{H}$ are then trivially given in terms of n by

$$E_n = \left(n + \frac{1}{2}\right)\hbar\omega$$

The proof can be broken down into the following steps:

1. $\langle n|\hat{a}^\dagger\hat{a}|n\rangle \geq 0 \implies n\langle n|n\rangle \geq 0 \implies n \geq 0$ since $|n\rangle$ is assumed to be a nontrivial (non-null) eigenstate of $\hat{a}^\dagger\hat{a}$. This implies that n has a lower bound, zero. We shall denote the lowest eigenvalue by n_{min}.

2. $[\hat{a},\hat{a}^\dagger] = 1 \implies \left[\hat{N},\hat{a}^\dagger\right] = \hat{a}^\dagger$ and $\left[\hat{N},\hat{a}\right] = -\hat{a}$.

3. $\left[\hat{N},\hat{a}^\dagger\right] = \hat{a}^\dagger \implies \hat{a}^\dagger|n\rangle = |n+1\rangle$ and $\left[\hat{N},\hat{a}\right] = -\hat{a} \implies \hat{a}|n\rangle = |n-1\rangle$.

4. $\hat{a}|n_{min}\rangle = |n_{min}-1\rangle \implies |n_{min}-1\rangle = |\,\rangle$. This is the only possible reconciliation if n_{min} is the lowest value of n.

5. $\hat{a}|n_{min}\rangle = |\,\rangle \implies \hat{a}^\dagger\hat{a}|n_{min}\rangle = |\,\rangle \implies n_{min}|n_{min}\rangle = |\,\rangle$. This implies that[5] $n_{min} = 0$.

6. By repeated application of $\hat{a}^\dagger$ on $|n_{min}\rangle$ it is evident that all nonnegative integers $n \in \mathbb{Z}_{0+}$ are eigenvalues of $\hat{N}$.

7. It only remains to show that *every* eigenvalue n must belong to $\mathbb{Z}_{0+}$. This follows because if $n \notin \mathbb{Z}_{0+}$, repeated action of $\hat{a}$ on $|n\rangle$ will eventually lead to $\hat{a}|n'\rangle = |\,\rangle$ for some n' within the range $0 < n' < 1$. But $\hat{a}|n'\rangle = |\,\rangle$ leads to $n' = 0$ as in step (5) leading to a contradiction.

Ground state

To determine the eigenstates, the basic strategy is to first determine the **ground state** (i.e., the lowest energy eigenstate) in some convenient basis (e.g., coordinate space) and then determine all the higher states by repeated application of $\hat{a}^\dagger$ on the ground state[6]. This *ground state* is given by the condition

$$\hat{a}|n_{min}\rangle = |\,\rangle$$

---

[5]If $c|\psi\rangle = |\,\rangle$ for some vector $|\psi\rangle$ and some scalar c, then either $c = 0$ or $|\psi\rangle = |\,\rangle$ (or both). Prove this.

[6]Note that the eigenvectors have already been (trivially) determined in the basis of eigenvectors of the number operator (i.e., in the number space).

We will cast this equation in the x representation

$$\langle x| \hat{a} |0\rangle = \langle x| \rangle$$

$$\langle x| \frac{1}{\sqrt{2}} \left(\sqrt{\frac{m\omega}{\hbar}} \hat{x} + \frac{i}{\sqrt{m\hbar\omega}} \hat{p} \right) |0\rangle = \langle x| \rangle$$

$$\left(\sqrt{\frac{m\omega}{\hbar}} x + \sqrt{\frac{\hbar}{m\omega}} \frac{d}{dx} \right) \langle x|0\rangle = 0$$

Writing $x_0 = \sqrt{\hbar/m\omega}$ and $\rho = x/x_0$, we have

$$\left(\frac{x}{x_0} + x_0 \frac{d}{dx} \right) \langle x|0\rangle = 0$$

$$\left(\rho + \frac{d}{d\rho} \right) \langle \rho|0\rangle = 0$$

The above differential equation is easily solved, yielding a *nondegenerate ground state*

$$\langle \rho|0\rangle = Ce^{-\frac{1}{2}\rho^2}$$

where the normalization constant C is given by

$$x_0 \int_{-\infty}^{+\infty} |\langle \rho|0\rangle|^2 d\rho = 1$$

$$C^2 x_0 \int_{-\infty}^{+\infty} e^{-\rho^2} d\rho = 1$$

so that we finally have

$$C = \sqrt{\frac{1}{x_0 \sqrt{\pi}}}$$

Degeneracy

We have already seen that the ground state is nondegenerate. It will be easy to check that all higher states will also be nondegenerate as a consequence. If $|n\rangle$ is a nondegenerate eigenstate, we have

$$\hat{a} |n+1; i\rangle = c^i_- |n\rangle$$

where i labels the possible degenerate states of the eigenvalue $n+1$ of $\hat{N}$ and c^i_- is some multiplicative constant. Hence

$$\hat{a}^\dagger \hat{a} |n+1; i\rangle = c^i_- \hat{a}^\dagger |n\rangle$$

$$(n+1) |n+1; i\rangle = c^i_- \hat{a}^\dagger |n\rangle$$

$$|n+1; i\rangle = \frac{c^i_-}{n+1} \left(\hat{a}^\dagger |n\rangle \right)$$

Thus all vectors $|n + 1; i\rangle$ turn out to be scalar multiples of $\hat{a}^\dagger |n\rangle$, and therefore cannot be independent. So the eigenvalue $n+1$ must also be nondegenerate if n is nondegenerate. Thus, by induction, all higher states must be nondegenerate.

Normalization constant

Since all the eigenstates are nondegenerate, the ladder action of $\hat{a}^\dagger$ and $\hat{a}$ implies

$$\hat{a}^\dagger |n\rangle = c_+^n |n + 1\rangle$$
$$\hat{a} |n\rangle = c_-^n |n - 1\rangle$$

where c_+^n and c_-^n are constants, and $|n\rangle, |n + 1\rangle$ and $|n - 1\rangle$ are all assumed to be normalized. The phases of all states $|n\rangle$ with respect to the ground state $|n = 0\rangle$ are by definition chosen in such a way that c_+^n and c_-^n are real and positive[7]. Taking self inner products on both sides we have

$$\langle n| \hat{a}\hat{a}^\dagger |n\rangle = |c_+^n|^2 \langle n + 1|n + 1\rangle$$
$$\langle n| \hat{a}^\dagger\hat{a} |n\rangle \quad |c_-^n|^2 \langle n - 1|n - 1\rangle$$

which gives

$$(n + 1) \langle n|n\rangle = |c_+^n|^2 \langle n + 1|n + 1\rangle$$
$$n \langle n|n\rangle = |c_-^n|^2 \langle n - 1|n - 1\rangle$$

where we have used the formulae, $\hat{N} = \hat{a}^\dagger\hat{a}$ and $[\hat{a}, \hat{a}^\dagger] = 1$. Hence we have

$$c_+^n = \sqrt{n + 1}$$
$$c_-^n = \sqrt{n}$$

Higher states

Now we can determine the eigenstates for $n > 1$. Since we have

$$\hat{a}^\dagger |n\rangle = \sqrt{n + 1} |n + 1\rangle$$

it is clear that all higher states can be constructed by repeated application of the operator $\hat{a}^\dagger$ on the ground state $|0\rangle$. This leads to

$$|n\rangle = \frac{1}{\sqrt{n!}} \left(\hat{a}^\dagger\right)^n |0\rangle$$

In the position representation, this equation becomes

$$\langle x|n\rangle = \langle x| \frac{1}{\sqrt{n!}} \left[\frac{1}{\sqrt{2}} \left(\sqrt{\frac{m\omega}{\hbar}}\hat{x} - \frac{i}{\sqrt{m\hbar\omega}}\hat{p}\right)\right]^n |0\rangle$$
$$\langle x|n\rangle = \frac{1}{\sqrt{n!}} \left[\frac{1}{\sqrt{2}} \left(\sqrt{\frac{m\omega}{\hbar}}x - \sqrt{\frac{\hbar}{m\omega}}\frac{d}{dx}\right)\right]^n \langle x|0\rangle$$

---

[7]Convince yourself that this can always be done.

In terms of the variable $\rho = x/x_0$ with $x_0 = \sqrt{\hbar/m\omega}$, we have

$$\langle \rho | n \rangle = \frac{1}{\sqrt{n!}} \left[\frac{1}{\sqrt{2}} \left(\rho - \frac{d}{d\rho} \right) \right]^n \langle \rho | 0 \rangle$$

This leads to[8]

$$\langle \rho | n \rangle = \sqrt{\frac{1}{x_0 \sqrt{\pi}\, 2^n\, n!}}\, H_n(\rho)\, e^{-\frac{1}{2}\rho^2}$$

where $H_n(\rho)$ are the Hermite polynomials of order n.

Problems

1. Show that eigenvalue equation for the Hamiltonian operator, cast in the momentum space, leads to the same differential equation as the one in position space.

2. For the n-th energy eigenstate of the harmonic oscillator

    (a) find out the expectation values $\langle \hat{x} \rangle$ and $\langle \hat{p} \rangle$ of the position and momentum observables.

    (Hint: Write the observables in terms of the ladder operators.)

    (b) obtain analytic expressions for the expectation values $\langle \hat{x}^2 \rangle$ and $\langle \hat{p}^2 \rangle$. Hence find the expressions for the expectation values of the kinetic and potential energies.

    (c) verify the uncertainty principle for the position and momentum observables.

3. By explicitly evaluating the integrals (in position space), compute the expectation values of the kinetic and potential energies for a harmonic oscillator in the second excited state ($n = 2$), and also verify the uncertainty principle for $\langle \hat{x} \rangle$ and $\langle \hat{p} \rangle$. Check that the findings are consistent with your general results obtained in the previous problem. You may use the standard integral:

$$2 \int_0^\infty x^{2n} e^{-\alpha x^2}\, dx = \frac{1.3.5 \ldots (2n-1)}{(2\alpha)^n} \sqrt{\pi/\alpha}$$

4. Imagine that a harmonic oscillator is in a superposition of energy eigenstates such that the probabilities of all the energies E are equal, and the relative phase between any two consecutive stationary states is θ.

---

[8]You will be asked to prove this result in the problem set that follows.

(a) What will be the probability that an imprecise measurement of energy yields $E \leq 2\hbar\omega$?

(b) Find out the probability that the oscillator can be found in $-\delta \leq x \leq +\delta$ (for some $\delta > 0$) after a time t has elapsed following the imprecise measurement in part (a) which yields $E \leq 2\hbar\omega$.

5. Consider a charged harmonic oscillator, having mass m and charge q, under the action of a uniform electric field E.

(a) Plot the potential by choosing the reference at the origin (i.e., $V|_{x=0} = 0$).

(b) Show that the effect of the electric field is to shift the energy levels by an amount $q^2 E^2 / 2m\omega^2$, and also obtain the modified energy eigenfunctions.

   (Hint: This practically requires no calculations.)

(c) An atom can sometimes be modelled as an electron bound harmonically to the nucleus. Show that an estimate of the atomic polarizability α is given by[9] $q^2/m\omega^2$. Does this expression agree with your physical expectation?

   (Hint: Compute the expectation value of the electric dipole moment in an eigenstate of energy. Imagine that the state moves from $|\psi_n\rangle$ to $|\psi_{n'}\rangle$ as the electric field is turned on. Here n and n' labels the original and shifted energies of the state of the electron.)

6. Show that

$$\left(\rho - \frac{d}{d\rho}\right)^n e^{-\frac{1}{2}\rho^2} = H_n(\rho) e^{-\frac{1}{2}\rho^2}$$

where $H_n(\rho)$ is the Hermite polynomial which was defined by the Rodrigues' formula.

(Hint: Use *method of induction*: Prove that the result is true for $n = 1$. Now assuming that it is true for $n = m$, prove that it is true for $n = m + 1$. Note that the case of $n = 0$ is trivially valid.)

7. If two operators $\hat{A}$ and $\hat{B}$ satisfy the commutation relation

$$\left[\hat{A}, \hat{B}\right] = b\hat{B}$$

for some real b, what can you say about the eigen-properties of $\hat{A}$?

---

[9]The atomic polarizability is the proportionality factor of the dipole moment and the electric field. It is the atomic parameter that determines the electric succeptibility.

8. We know that the energy eigenfunctions of the harmonic oscillator must be orthogonal (since the eigenvalues are nondegenerate). In fact, they form an orthonormal basis in $\mathbb{L}^2(\mathbb{R})$. Using this fact

   (a) modify the inner product such that the Hermite polynomials may be called orthogonal. Verify explicitly using the definition of Hermite polynomials (Rodrigues' formula) that they indeed obey these orthogonality conditions.

   (b) suggest the condition on the set of functions for which the Hermite polynomials will be *complete*.

Angular Momentum

We now wish to describe a set of quantum observables that is not tied to any particular system as such but play a very important role in the description of many quantum systems. These observables originate as quantum transcriptions of classical orbital angular momenta[10].

Orbital angular momentum

Consider a particle moving in three dimensions. The angular momentum $\vec{L}$ of the particle having position $\vec{r}$ and moving with a momentum $\vec{p}$ is defined by

$$\vec{L} = \vec{r} \times \vec{p}$$

In QM, orbital angular momenta comprises a set of three operators, $\hat{L}_i$ ($i = 1, 2, 3$) corresponding to the x, y and z directions respectively, defined by

$$\hat{L}_i = \sum_{j,k=1}^{3} \epsilon_{ijk} \hat{x}_j \hat{p}_k$$

where $\hat{x}_j$ and $\hat{p}_k$ are the position and momentum operators, and ϵ_{ijk} is the *Levi-Civita* symbol defined by

$$\epsilon_{ijk} = \begin{cases} +1 & \text{for } i=1, j=2, k=3 \text{ and cyclic permutations} \\ -1 & \text{for } i=3, j=2, k=1 \text{ and cyclic permutations} \\ 0 & \text{when two or more indices are repeated} \end{cases}$$

---

[10]As in classical mechanics, the angular momenta observables also play a pivotal role in the description of *rotational symmetry* in QM. Unfortunately, we shall not be discussing *symmetries* in this book. For a more exhaustive treatment of angular momentum in QM, see (Cohen-Tannoudji et al. 1977). In particular, see *complement B* of chapter six for its connection with rotational symmetry.

One also defines the "square of the total angular momentum" operator $\hat{L}^2$ by

$$\hat{L}^2 = \sum_{i=1}^{3} \hat{L}_i^2$$

Starting from the canonical commutation relations $[\hat{x}_i, \hat{p}_j] = i\hbar\delta_{ij}$, or the explicit representations of $\hat{x}_i$ and $\hat{p}_j$ given by $[\hat{x}_i] = x_i$ and $[\hat{p}_i] = -i\hbar\,(\partial/\partial x_i)$, it is easy to check that

$$\left[\hat{L}_i, \hat{L}_j\right] = i\hbar\epsilon_{ijk}\hat{L}_k$$

From these commutation relations, it follows that

$$\left[\hat{L}^2, \hat{L}_i\right] = 0$$

Thus, a complete set of common eigenvectors of $\hat{L}^2$ and $\hat{L}_i$ will exist. Here $\hat{L}_i$ is an arbitrarily chosen angular momentum operator from the set of the three angular momenta. It is customary to discuss the eigenvalues and common eigenvectors of $\hat{L}^2$ and $\hat{L}_z$. This is what we shall do now.

We write the eigenvalue equations for $\hat{L}^2$ and $\hat{L}_z$ as

$$\begin{aligned}
\hat{L}^2 \left|\lambda, m, k\right\rangle &= \hbar^2\lambda \left|\lambda, m, k\right\rangle \\
\hat{L}_z \left|\lambda, m, k\right\rangle &= \hbar m \left|\lambda, m, k\right\rangle
\end{aligned}$$

where λ and m are now dimensionless numbers and k labels any possible degeneracy. For algebraic convenience we write[11]

$$\lambda = l\,(l+1)$$

so that the eigenvalue equation for $\hat{L}^2$ becomes

$$\hat{L}^2 \left|l, m, k\right\rangle = \hbar^2 l\,(l+1) \left|l, m, k\right\rangle$$

For historical reasons, l is called the ***azimuthal quantum number***, and m is called the ***magnetic quantum number***. Starting from the representations of $\hat{x}_i$ and $\hat{p}_j$, one can easily derive the (x, y, z) space representations of $\hat{L}^2$ and $\hat{L}_i$. Thereafter, by a simple change of variables, we can derive the (r, θ, ϕ) space representations of $\hat{L}^2$ and $\hat{L}_i$. These representations for $\hat{L}^2$ and $\hat{L}_z$ are

$$\langle r, \theta, \phi|\, \hat{L}^2 = -\frac{\hbar^2}{\sin^2\theta}\left[\sin\theta\frac{\partial}{\partial\theta}\left(\sin\theta\frac{\partial}{\partial\theta}\right) + \frac{\partial^2}{\partial\phi^2}\right]\langle r, \theta, \phi|$$

$$\langle r, \theta, \phi|\, \hat{L}_z = -i\hbar\frac{\partial}{\partial\phi}\langle r, \theta, \phi|$$

---

[11]This is a quadratic equation in l implying that there are two roots of l for each value of λ. However, it is easy to see that λ is nonnegative, and one can check that there is exactly one positive and one negative root for each positive value of λ. So there is a one to one mapping between the positive roots of l and the positive eigenvalues λ. Using $l = 0$ to correspond to $\lambda = 0$ one can use the nonnegative roots of l to label the eigenvalues λ.

The eigenvalue equations for $\hat{L}^2$ and $\hat{L}_z$, cast in the (r, θ, ϕ) space, become the following differential equations:

$$-\frac{\hbar^2}{\sin^2\theta}\left[\sin\theta\frac{\partial}{\partial\theta}\left(\sin\theta\frac{\partial}{\partial\theta}\right) + \frac{\partial^2}{\partial\phi^2}\right]\langle r,\theta,\phi\,|l,m,k\rangle = \hbar^2 l\,(l+1)\,\langle r,\theta,\phi\,|l,m,k\rangle$$

$$-i\hbar\frac{\partial}{\partial\phi}\langle r,\theta,\phi\,|l,m,k\rangle = \hbar m\,\langle r,\theta,\phi\,|l,m,k\rangle$$

Let us write the solutions as

$$\langle r,\theta,\phi\,|l,m,k\rangle = f_k\,(r)\,Y_l^m\,(\theta,\phi)$$

where the $f_k(r)$ are arbitrary (square integrable) multiplicative functions of[12] r.

Now to solve these equations, one can use the standard method of separation of variables. Thus we write

$$Y_l^m\,(\theta,\phi) = \Theta\,(\theta)\,\Phi\,(\phi)$$

(where we have suppressed the indices l and m in $\Theta\,(\theta)$ and $\Phi\,(\phi)$).

The eigenvalue equation for $\hat{L}_z$ then becomes

$$-i\hbar\frac{d\Phi\,(\phi)}{d\phi} = \hbar m\Phi\,(\phi)$$

which has a simple solution

$$\Phi\,(\phi) = Ce^{im\phi}$$

where C is a constant of normalization which we shall henceforth absorb in $\Theta\,(\theta)$. In order to ensure that $\Phi\,(\phi + 2\pi) = \Phi\,(\phi)$, the index m must be integral.

The eigenvalue equation for the operator $\hat{L}^2$ now becomes the ordinary differential equation[13]

$$\frac{1}{\sin\theta}\frac{d}{d\theta}\left(\sin\theta\frac{d\Theta}{d\theta}\right) + \left[l\,(l+1) - \frac{m^2}{\sin^2\theta}\right]\Theta = 0$$

---

[12]Since the representations of $\hat{L}^2$ and $\hat{L}_i$ are independent of r, any solution to the differential equations are essentially functions of θ and ϕ. In the (r,θ,ϕ) space, this means that $Y_l^m\,(\theta,\phi)$ are unique up to a multiplicative function of r which must, of course, also be square integrable on physical grounds.

[13]One could also start from the $\hat{L}^2$ eigenvalue equation whence, employing separation of variables (with separation constant m^2), one would get a ϕ - differential equation

$$\frac{1}{\Phi}\frac{d^2\Phi}{d\phi^2} = -m^2$$

This leads to a solution which is automatically obeyed by the eigenvalue equation for $\hat{L}_z$.

Setting $u = \cos\theta$, we have

$$\left[\frac{d}{du}\left(1 - u^2\right)\frac{d\Theta}{du}\right] + \left[l\left(l + 1\right) - \frac{m^2}{1 - u^2}\right]\Theta = 0$$

with $-1 \leq u \leq +1$. This is a standard, well studied differential equation that is usually discussed in connection with Legendre's differential equation; see, for example (Margenau and Murphy 1956). The solutions to this equation are given by the so called ***associated Legendre polynomials***[14] $P_l^m(u)$:

$$P_l^m(u) = \left(1 - u^2\right)^{\frac{|m|}{2}}\left(\frac{d}{du}\right)^{|m|} P_l(u) \text{ with } P_l(u) = \frac{(-1)^l}{2^l l!}\left(\frac{d}{du}\right)^l \left(1 - u^2\right)^l$$

where $P_l(u)$ are called the ***Legendre polynomials***. They are well behaved in $-1 \leq u \leq +1$ (i.e., bounded, square integrable, etc.) for nonnegative integral values of[15] l. It is also clear from the defining equation of $P_l^m(u)$, that for $P_l^m(u)$ to be nonzero, $|m| \leq l$.

Finally, after normalization, putting everything together, we have

$$Y_l^m(\theta, \phi) = \eta\sqrt{\frac{(2l + 1)}{4\pi}\frac{(l - |m|)!}{(l + |m|)!}} P_l^m(\cos\theta) e^{im\phi}$$

where in order to be consistent with standard phase conventions (that remain undecided after normalization), we choose $\eta = (-1)^m$ for $m \geq 0$ and $\eta = 1$ when[16] $m < 0$. Here $l \in \{0, 1, 2, \ldots\}$ and $m \in \{-l, -l + 1, \ldots, l\}$, as explained before, for acceptable well behaved solutions (which are finite, single valued and square integrable). These important functions $Y_l^m(\theta, \phi)$ are known in the literature as ***spherical harmonics***.

Since r has dropped out from the representations of $\hat{L}^2$ and $\hat{L}_z$, it also means that we could just as well have defined the angular momenta observables in the (θ, ϕ) space. It is not hard to imagine a physical system for which such a space would be appropriate: a particle moving on a spherical surface of some radius a. The position of such a particle can obviously be described by the polar and azimuthal angles θ and ϕ respectively. The quantum states of this system would live in the space of all square integrable functions of θ and ϕ, which in turn will be contained in a vector space that can be formally written as the linear span of $\{|\theta, \phi\rangle : \theta \in [0, \pi]\ , \phi \in [0, 2\pi]\}$. Here $|\theta, \phi\rangle$ are, obviously, the common eigenvectors of the observables $\hat{\theta}$ and $\hat{\phi}$ which are assumed to form

---

[14]Note that for $m = 0$, if we set $l(l + 1) = \lambda$, the differential equation actually reduces to Legendre's differential equation, whose solutions are the Legendre polynomials. Reverting to $\lambda = l(l + 1)$ and differentiating the Legendre equation m times (with $\Theta = P_l(u)$) shows that the associated Legendre polynomials are indeed the solution to the given equation.

[15]The defining equation for the Legendre polynomial (Rodrigues' formula) does not even make sense if l is not a nonnegative integer.

[16]We shall review this more clearly, later in the chapter.

a CSCO for this space[17] The angular momentum operators $\hat{L}^2$ and $\hat{L}_z$ also form a CSCO for this space. The inner product in this space is defined as

$$\langle \beta | \alpha \rangle = \int_{\theta=0}^{\pi} \int_{\phi=0}^{2\pi} \beta^* \left(\theta, \phi\right) \alpha \left(\theta, \phi\right) \sin \theta \, d\theta \, d\phi$$

where the appropriate measure is provided by the volume element $\sin \theta \, d\theta \, d\phi$. By the same token in the (r, θ, ϕ) space the inner product is given by

$$\langle \beta | \alpha \rangle = \int_{r=0}^{\infty} \int_{\theta=0}^{\pi} \int_{\phi=0}^{2\pi} \beta^* \left(r, \theta, \phi\right) \alpha \left(r, \theta, \phi\right) r^2 \sin \theta \, dr \, d\theta \, d\phi$$

Here the appropriate volume element is $r^2 \sin \theta \, d\theta \, d\phi$. Incidentally, for a similar reason the requirement that $f_k \left(r\right)$ must be square integrable means that $\int_0^{\infty} f_k \left(r\right) r^2 dr$ should be finite, not $\int_0^{\infty} f_k(r) \, dr$.

Generalized angular momentum

In QM, the idea of angular momenta is not restricted to the orbital angular momenta that we have discussed above. Instead, the concept is defined by abstracting and generalizing the *angular momentum commutation relations*:

$$\left[\hat{J}_i, \hat{J}_j\right] \;=\; i\hbar \epsilon_{ijk} \hat{J}_k$$

Thus, any set of three observables $\left\{\hat{J}_i \, ; \, i = 1, 2, 3\right\}$ satisfying the above commutation relations are called angular momenta in QM. But why does one care about such sets of observables? Well, it turns out that the description of a huge body of fundamental quantum observables, which do not have classical analogues, is based on or inspired by the mathematical framework of generalized angular momenta[18]. Unfortunately, we shall not be able to go into all that in this book. In this section we shall simply show how the algebraic method, that was demonstrated for the harmonic oscillator, can be used to solve the eigenvalue problem for the generalized angular momenta.

Eigenvalue spectrum

First of all, we define an operator called *"total angular momentum squared"* by

$$\hat{J}^2 \;=\; \sum_{i=1}^{3} \hat{J}_i^2$$

---

[17]the linear span of $\{|\theta, \phi\rangle : \theta \in [0, \pi], \, \phi \in [0, 2\pi]\}$ is obviously identical to the linear span of $\left\{|x, y, z\rangle : x^2 + y^2 + z^2 = a^2\right\}$ which in turn is embedded in the linear span of $\{|x, y, z\rangle : x, y, z \in (-\infty, +\infty)\}$.

[18]Historically, the idea of generalized angular momenta has also had a profound influence on the formulation of QM as we know it today.

It is easy to check that $\hat{J}^2$ commutes with all $\hat{J}_i$'s:

$$\left[\hat{J}^2,\, \hat{J}_i\right] = 0$$

The eigenvalue equations for $\hat{J}^2$ and $\hat{J}_3$ are[19]

$$\begin{array}{rcl} \hat{J}^2\,|\lambda, m\rangle & = & \hbar^2\lambda\,|\lambda, m\rangle \\ \hat{J}_3\,|\lambda, m\rangle & = & \hbar m\,|\lambda, m\rangle \end{array}$$

For algebraic convenience (just as we had done before), we write

$$\lambda = j\,(j+1)$$

We shall prove that

$$j = \frac{N}{2} \quad \text{where} \quad N \in \{0, 1, 2, \ldots\} \quad \text{and} \quad m \in \{-j, -j+1, \ldots, +j\}$$

We define *ladder operators*, $\hat{J}_+$ and $\hat{J}_-$, by

$$\begin{array}{rcl} \hat{J}_+ & = & \hat{J}_1 + i\,\hat{J}_2 \\ \hat{J}_- & = & \hat{J}_1 - i\,\hat{J}_2 \end{array}$$

Evidently, $\hat{J}_+$ and $\hat{J}_-$ are Hermitian conjugates of each other: $\hat{J}_+ = \hat{J}_-^\dagger$

It is easy to check the commutation relations

$$\begin{array}{rcl} \left[\hat{J}_3,\, \hat{J}_+\right] & = & \hbar\hat{J}_+ \\ \left[\hat{J}_3,\, \hat{J}_-\right] & = & -\hbar\hat{J}_- \end{array}$$

and the operator relations

$$\begin{array}{rcl} \hat{J}_+\hat{J}_- & = & \hat{J}^2 - \hat{J}_3^2 + \hbar\hat{J}_3 \\ \hat{J}_-\hat{J}_+ & = & \hat{J}^2 - \hat{J}_3^2 - \hbar\hat{J}_3 \end{array}$$

The main argument of the proof comprises the following steps[20]:

1. The parameter j can be taken to be nonnegative[21].

---

[19] Although the notation $|\lambda, m\rangle$ for the eigenvectors seem to suggest that $\hat{J}^2$ and $\hat{J}_3$ forms a CSCO so that λ and m specifies a unique common eigenvector, we are actually not making any such assumption. The issue of completeness of the operators will arise only when we specify the system with which the generalized angular momenta would be associated.

[20] In what follows, the line of reasoning is very similar to what was done for the harmonic oscillator.

[21] We have explained this before when we set $\lambda = l\,(l+1)$ in the previous subsection.

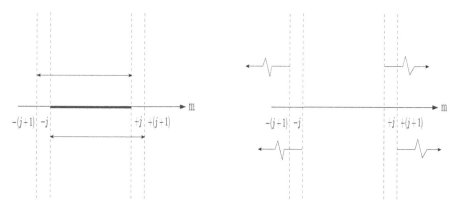

FIGURE 8.1: Range of allowed 'm' values.

2. $\hat{J}_+ = \hat{J}_-^\dagger \implies \langle j, m | \hat{J}_- \hat{J}_+ | j, m \rangle \geq 0$. But $\hat{J}_- \hat{J}_+ = \hat{J}^2 - \hat{J}_3^2 - \hbar \hat{J}_3$
   $\implies j(j+1) - m(m+1) \geq 0 \implies (j-m)(j+m+1) \geq 0$
   $\implies m \leq j$ and $m \geq -(j+1)$ OR $m \geq j$ and $m \leq -(j+1)$
   The second case is impossible.

3. $\hat{J}_+ = \hat{J}_-^\dagger \implies \langle j, m | \hat{J}_+ \hat{J}_- | j, m \rangle \geq 0$. But $\hat{J}_+ \hat{J}_- = \hat{J}^2 - \hat{J}_3^2 + \hbar \hat{J}_3$
   $\implies j(j+1) - m(m-1) \geq 0 \implies (j+m)(j-m+1) \geq 0$
   $\implies m \geq -j$ and $m \leq j+1$ OR $m \leq -j$ and $m \geq j+1$
   The second case is impossible.

4. From (2) and (3) it follows that $-j \leq m \leq +j$ (see Figure 8.1). Thus, there will exist a maximum value m_{max}, and a minimum value m_{min} of m bounded by $+j$ and $-j$ respectively.

5. $\left[\hat{J}_3, \hat{J}_+\right] = \hbar \hat{J}_+$ and $\left[\hat{J}_3, \hat{J}_-\right] = -\hbar \hat{J}_-$
   $\implies \hat{J}_+ | j, m \rangle = | j, m+1 \rangle$ and $\hat{J}_- | j, m \rangle = | j, m-1 \rangle$

6. Since m is bounded above for given j,
   $\hat{J}_+ | j, m_{max} \rangle = | \rangle \implies \hat{J}_- \hat{J}_+ | j, m_{max} \rangle = | \rangle$
   $\implies (j - m_{max})(j + m_{max} + 1) | j, m_{max} \rangle = | \rangle \implies j = m_{max}$

7. Similarly, since m is bounded below,
   $\hat{J}_- | j, m_{min} \rangle = | \rangle \implies j = m_{min}$

8. Therefore $m \in \{-j, -j+1, \ldots, +j\}$.

9. Intermediate values are not possible because repeated action of ladder operators will take them beyond the allowed range. For consistency, we must then require $\hat{J}_- | j, m' \rangle = | \rangle$ and $\hat{J}_+ | j, m'' \rangle = | \rangle$ for some m' and m'' satisfying $m_{min} < m' < m_{min} + 1$ and $m_{max} - 1 < m'' < m_{max}$. But just as in points (7) and (6), we have $\hat{J}_- | j, m' \rangle = | \rangle \implies m' = m_{min}$ and $\hat{J}_+ | j, m'' \rangle = | \rangle \implies m'' = m_{max}$ leading to contradictions.

10. Finally, since there are $2j+1$ allowed values of m, $2j+1$ must be a natural number. Hence j must be half integral: $j = N/2$ with $N \in \{0, 1, 2, \ldots\}$.

Degeneracy

So far we have been using the misleading symbol $|j, m\rangle$ for the common eigenvectors of $\hat{J}^2$ and $\hat{J}_3$ which seems to suggest that $\left\{\hat{J}^2, \hat{J}_3\right\}$ is a CSCO. We had, however, pointed out (in a footnote) that this does not necessarily have to be the case. To discuss degeneracy, let us denote the eigenvectors as $|j, m, p\rangle$ where the additional index p is used to label all the independent eigenvectors associated with a particular set (j, m). For concreteness, we can assume that $\left\{\hat{J}^2, \hat{J}_3, \hat{P}\right\}$ forms a CSCO so that p labels the eigenvalues of $\hat{P}$ (this can actually be done without any loss of generality). Let us assume that there are $g(j, m)$ independent eigenvectors associated with the set (j, m). It is not hard to show that $g(j, m)$ is actually independent of m so that one can actually write[22] $g(j, m) = g(j)$.

The degree of degeneracy of the eigenvalues of $\hat{J}^2$ and $\hat{J}_3$ are then easy to see:

1. To every value of j there are $(2j + 1) \times g(j)$ independent eigenvectors $|j, m, p\rangle$.

2. If for a given value of m there are r values of j such that $|m| \leq j$ (the allowed values of j depends on the actual system being described), then the number of independent eigenvectors associated with this value of m are $r \times g(j)$.

We can imagine a subspace $V^{(j,m)}$ associated with the set (j, m), which is a linear span of an orthonormal basis $\{|j, m, p\rangle \, ; \, p = 1, 2, \ldots, g(j)\}$. The full inner product space describing the system can be split up into such subspaces $V^{(j,-j)}, V^{(j,-j+1)}, \ldots, V^{(j,j)}$ where the subspace $V^{(j,m+1)}$ is the linear span of the orthonormal basis $\left\{\hat{J}_+ |j, m, p\rangle \, ; \, p = 1, 2, \ldots, g(j)\right\}$. Of course, one needs to prove that this set actually forms an orthonormal basis. It is not hard to show this, and I leave it as an exercise[23].

Eigenvectors

To obtain the eigenvectors explicitly we may first cast the equation

$$\hat{J}_+ |j, j\rangle = | \ \rangle$$

in the representation of our choice, and then all the other states $|j, m\rangle$ with

---

[22]See (Cohen-Tannoudji et al. 1977).

[23]Incidentally, in this case one says that the full inner product space, say V, is a *direct sum* of the subspaces $V^{(j,-j)}, V^{(j,-j+1)}, \ldots, V^{(j,j)}$. See appendix 'B' for the definition of *direct sum*.

$m = -j, -j+1, \ldots, j-1$ can be explicitly obtained by repeated application of[24]

$$\hat{J}_- \,|j, m\rangle = |j, m-1\rangle$$

Normalization constants

The ladder action of $\hat{J}_\pm$ implies

$$\begin{aligned}
\hat{J}_+ \,|j, m\rangle &= c_+ \,|j, m+1\rangle \\
\hat{J}_- \,|j, m\rangle &= c_- \,|j, m-1\rangle
\end{aligned}$$

where $|j, m\rangle$, $|j, m+1\rangle$ and $|j, m-1\rangle$ are assumed normalized, and correspondingly, c_+ and c_- are appropriate constants. The constants c_+ and c_- can be taken to be real by making appropriate choice of the phases of the vectors $|j, m\rangle$. Taking self inner products on both sides and using the expressions for $\hat{J}_- \hat{J}_+$ and $\hat{J}_+ \hat{J}_-$ that were given earlier, we have

$$\begin{aligned}
\hbar^2 \left(j(j+1) - m(m+1)\right) \langle j, m|j, m\rangle &= c_+^2 \,\langle j, m+1|j, m+1\rangle \\
\hbar^2 \left(j(j+1) - m(m-1)\right) \langle j, m|j, m\rangle &= c_-^2 \,\langle j, m-1|j, m-1\rangle
\end{aligned}$$

which gives

$$\begin{aligned}
c_+ &= \hbar\sqrt{j(j+1) - m(m+1)} \\
c_- &= \hbar\sqrt{j(j+1) - m(m-1)}
\end{aligned}$$

Thus, finally, the precise action of the ladder operators can be written as

$$\begin{aligned}
\hat{J}_+ \,|j, m\rangle &= \hbar\sqrt{j(j+1) - m(m+1)} \,|j, m+1\rangle \\
\hat{J}_- \,|j, m\rangle &= \hbar\sqrt{j(j+1) - m(m-1)} \,|j, m-1\rangle
\end{aligned}$$

Special case of orbital angular momenta

The orbital angular momenta are, obviously, a special case of generalized angular momenta. Using the ladder operator technique discussed above, we can carry out the construction of the eigenvectors explicitly in the (θ, ϕ) space for the case of orbital angular momenta. Let us use the alphabet 'L' instead of 'J' to denote the orbital angular momentum operators, and l in place of 'j' to label the eigenvalues of the total angular momentum squared operator. We can start by casting the equation

$$\hat{L}_+ \,|l, l\rangle = |\ \rangle$$

in the (θ, ϕ) space

$$\langle \theta, \phi| \,\hat{L}_+ \,|l, l\rangle = \langle \theta, \phi \,|\ \rangle = 0$$

---

[24]Of course, we could also start with $\hat{J}_- \,|j, -j\rangle = |\ \rangle$ and then generate all the other states by repeated application of $\hat{J}_+$ on $|j, -j\rangle$.

All the other states $|l, m\rangle$ with $-l \leq m < +l$ can be obtained by repeated use of $\hat{L}_-$:

$$\langle \theta, \phi| \hat{L}_- |l, m\rangle = c_- \langle \theta, \phi |l, m - 1\rangle$$

We have already seen the (θ, ϕ) representations of $\hat{L}^2$ and $\hat{L}_z$. The representations of $\hat{L}_x$ and $\hat{L}_y$ are given by

$$\langle \theta, \phi| \hat{L}_x = i\hbar \left[\sin \phi \frac{\partial}{\partial \theta} + \cos \phi \cot \theta \frac{\partial}{\partial \phi} \right] \langle \theta, \phi|$$

$$\langle \theta, \phi| \hat{L}_y = -i\hbar \left[-\cos \phi \frac{\partial}{\partial \theta} + \sin \phi \cot \theta \frac{\partial}{\partial \phi} \right] \langle \theta, \phi|$$

Hence, the representations of the ladder operators, $\hat{L}_+$ and $\hat{L}_-$, will be given by

$$\langle \theta, \phi| \hat{L}_\pm = \pm\hbar e^{\pm i\phi} \left[\frac{\partial}{\partial \theta} \pm i \cot \theta \frac{\partial}{\partial \phi} \right] \langle \theta, \phi|$$

Now the eigenvalue equation for $\hat{L}_z$ in the (θ, ϕ) space is

$$-i\hbar \frac{\partial Y_l^m (\theta, \phi)}{\partial \phi} = \hbar m Y_l^m (\theta, \phi)$$

This equation is solved trivially:

$$Y_l^m (\theta, \phi) = F_{l,m} (\theta) e^{im\phi}$$

where $F_{l,m} (\theta)$ is some function of θ. Requiring the natural periodicity $e^{im\phi} = e^{im(\phi+2\pi)}$, we have $|m| = 0, 1, 2, \ldots$ which in turn requires that $l = 0, 1, 2, \ldots$. Note that, in this case, the single valuedness of the eigenfunction imposes a further restriction on the spectra of the operators $\hat{L}^2$ and $\hat{L}_z$ (forbidding half-integral values).

Now we have

$$\langle \theta, \phi| \hat{L}_+ |l, l\rangle = 0$$
$$\hbar e^{i\phi} \left[\frac{\partial}{\partial \theta} + i \cot \theta \frac{\partial}{\partial \phi} \right] Y_l^l (\theta, \phi) = 0$$
$$\left[\frac{d}{d\theta} - l \cot \theta \right] F_{l,l} (\theta) = 0$$

leading to an easy solution

$$F_{l,l} (\theta) = A_l (\sin \theta)^l$$

so that

$$Y_l^l (\theta, \phi) = A_l (\sin \theta)^l e^{il\phi}$$

Here A_l is a constant of normalization. Now for a fixed l, all the other states $|l, l-1\rangle, |l, l-2\rangle, \ldots, |l, -l\rangle$ in the (θ, ϕ) space will be obtained by repeated action of $\hat{L}_-$ on $Y_l^l(\theta, \phi)$. Upto normalization we have

$$Y_l^{l-1}(\theta, \phi) = -\hbar e^{-i\phi}\left[\frac{\partial}{\partial\theta} - i\cot\theta\frac{\partial}{\partial\phi}\right]Y_l^l(\theta, \phi)$$

$$Y_l^{l-2}(\theta, \phi) = -\hbar e^{-i\phi}\left[\frac{\partial}{\partial\theta} - i\cot\theta\frac{\partial}{\partial\phi}\right]Y_l^{l-1}(\theta, \phi)$$

$$\ldots = \ldots$$

It is evident that the eigenvector $|l, l\rangle$ is unique up to the normalization constant A_l. So using the same reasoning that was used to prove the degeneracy of the eigenstates of the harmonic oscillator, it should be easy to show that the uniqueness of the state $|l, l\rangle$ implies that all states $|l, l-1\rangle, |l, l-2\rangle, \ldots, |l, -l\rangle$ are unique up to a multiplicative constant.

Problems

1. Measurement of the total angular momentum squared operator $\hat{J}^2$ on a certain system leads to the eigenvalue $2\hbar^2$ and the state collapses to the appropriate subspace. What would be the possible outcomes of an instantaneously subsequent measurement of $\hat{J}_1$.

2. Consider a state $|\psi\rangle$ in a linear superposition of the angular momentum states $|j, m\rangle$ (common eigenstates of $\hat{J}^2$ and $\hat{J}_3$) given by

$$|\psi\rangle = |2, 2\rangle \langle 2, 2|\psi\rangle + |2, 1\rangle \langle 2, 1|\psi\rangle + |2, 0\rangle \langle 2, 0|\psi\rangle$$
$$+ |1, 1\rangle \langle 1, 1|\psi\rangle + |1, -1\rangle \langle 1, -1|\psi\rangle$$

   (a) What would be the final state if a measurement of $\hat{J}^2$ yields $6\hbar^2$?
   (b) If a measurement of the operator $\hat{J}_1^2 + \hat{J}_2^2$ is now made, what are the expected outcomes and their probabilities?

3. A system is in a common eigenstate $|j, m\rangle$ of the angular momentum operators $\hat{J}^2$ and $\hat{J}_3$ with respective eigenvalues $j(j+1)\hbar^2$ and $m\hbar$. Calculate the expectation values $\left\langle \hat{J}_1 \right\rangle$ and $\left\langle \hat{J}_1^2 \right\rangle$.

4. Find out the representation of the set of angular momentum operators $\left\{\hat{J}_+, \hat{J}_-, \hat{J}_1, \hat{J}_2, \hat{J}_3, \hat{J}^2\right\}$ in a basis of common eigenstates of the operators $\hat{J}^2$ and $\hat{J}_3$ for i) $j = 3/2$ and ii) $j = 1$.

   $\Big($Hint: Start by working out the representations of the ladder operators and express the other operators in terms of them.$\Big)$

5. If $\hat{J}_1$ and $\hat{J}_2$ are generalized angular momentum operators, check whether the operators $\hat{A}$ and $\hat{B}$ are compatible, where

$$\hat{A} = \hat{J}_1 + \hat{J}_2 \quad \text{and} \quad \hat{B} = \hat{J}_1 - \hat{J}_2$$

6. Consider the set of three operators $\left\{\hat{K}_1, \hat{K}_2, \hat{K}_3\right\}$ obeying the commutation relations

$$\left[\hat{K}_+, \hat{K}_-\right] = 2\hbar\hat{K}_3 \qquad \left[\hat{K}_\pm, \hat{K}_3\right] = \mp\hbar\hat{K}_\pm$$

where $\hat{K}_\pm$ are the ladder operators defined in the usual way. Show that these commutation relations define an equivalent way to specify that the operators $\left\{\hat{K}_1, \hat{K}_2, \hat{K}_3\right\}$ are angular momenta.

7. Classically, the magnetic moment $\vec{M}$ of a charged particle is proportional to its angular momentum $\vec{L}$:

$$\vec{M} = \gamma\vec{L}$$

where γ is the proportionality constant called the gyromagnetic ratio. Consider a point mass (which, therefore, cannot spin) moving with zero orbital angular momentum. Imagine that, strangely, this system yields a magnetic moment spectrum comprising two values, say μ and $-\mu$. Postulate an *intrinsic generalized angular momentum* with spectrum $(\hbar/2, -\hbar/2)$, which, as a figure of speech, you might call *spin*, that produces the magnetic moments with: $\mu = \bar{\gamma}\hbar/2$ ($\bar{\gamma}$: some new constant).

   (a) Argue that the total spin s must be $1/2$ for the particle.
       (here s is the value of the quantum number that labels the eigenvalue for total angular momentum squared which we called j earlier)

   (b) For some chosen axis, say $\hat{z}$, let the members of the spectrum of the spin $\hat{S}_z$ be $(\hbar/2, -\hbar/2)$. Construct the operators $\hat{S}^2$ and $\hat{S}_z$ in a basis of their common eigenstates.

   (c) Now construct representations for the ladder operators and hence construct the representations for $\hat{S}_x$ and $\hat{S}_y$.

   (d) Verify that $\hat{S}_x$, $\hat{S}_y$ and $\hat{S}_z$ does indeed satisfy the angular momentum commutation relations.

   (e) If the magnetic moment spectrum comprised three equispaced values μ, 0 and $-\mu$ instead of two, what would be the total spin? Repeat the entire exercise for this case.

8. Derive the momentum space representations of the orbital angular momenta $\left\{\hat{L}_x, \hat{L}_y, \hat{L}_z\right\}$.

9. Using the canonical commutation relations, check whether the following set of observables are compatible:

$$i) \ \left\{\hat{y}, \hat{L}_y\right\} \quad \text{and} \quad i) \ \left\{\hat{y}, \hat{L}_z\right\}$$

Here $\left\{\hat{L}_x, \hat{L}_y, \hat{L}_z\right\}$ are orbital angular momenta and $\hat{y}$ is the usual position operator along the y-axis.

10. Given that the spherical harmonics form an orthonormal basis in the (θ, ϕ) space, argue that

$$\sum_{l=0}^{\infty} \sum_{m=-l}^{l} Y_l^m(\theta, \phi) Y_l^{m*}(\theta', \phi') = \delta(\cos\theta - \cos\theta')\delta(\cos\phi - \cos\phi')$$

11. Derive the explicit formulae for the spherical harmonics by making use of the repeated action of the ladder operators on Y_l^l and Y_l^{-l}. You may proceed through the following exercises:

(a) Using $\hat{L}_+ Y_l^l = 0$, show that the normalized form of Y_l^l is given by

$$Y_l^l(\theta, \phi) = \frac{(-1)^l}{2^l l!} \sqrt{\frac{(2l+1)!}{4\pi}} (\sin\theta)^l e^{il\phi}$$

with the $(-1)^l$ being a customary choice of phase (that is not decided by normalization[25]).

(b) Prove that

$$\hat{L}_\pm^k \left[f(\theta)e^{im\phi}\right] = (\pm\hbar)^k e^{i(m\pm k)\phi} (\sin\theta)^{k\pm m} \frac{d^k}{d(\cos\theta)^k} \left[(\sin\theta)^{\mp m} f(\theta)\right]$$

(Hint: Use method of induction. First prove that it is true for $k=1$. Then show using this formula that if it true for $k = s-1$, then it is true for $k = s$.)

(c) Use the standard lowering ladder operator action to show that

$$Y_l^m(\theta, \phi) = \sqrt{\frac{(l+m)!}{2l!(l-m)!}} \left(\frac{\hat{L}_-}{\hbar}\right)^{l-m} Y_l^l(\theta, \phi)$$

from which the desired expression follows by applying the formula in part (b):

$$Y_l^m(\theta, \phi) = \frac{(-1)^l}{2^l l!} \sqrt{\frac{(2l+1)}{4\pi} \frac{(l+m)!}{(l-m)!}} e^{im\phi} (\sin\theta)^{-m} \frac{d^{l-m}}{d(\cos\theta)^{l-m}} (\sin\theta)^{2l}$$

(d) Now from the above expression, show that

$$Y_l^{-l}(\theta, \phi) = \frac{1}{2^l l!} \sqrt{\frac{(2l+1)!}{4\pi}} (\sin\theta)^l e^{-il\phi}$$

---

[25] This phase convention ensures that $Y_l^0(\theta, \phi)$ (which is ϕ-independent) is real and positive.

and hence demonstrate that an alternative (but completely equivalent) representation of $Y_l^m(\theta, \phi)$ constructed using $l + m$ applications of the raising operator $\hat{L}_+$ on $Y_l^{-l}(\theta, \phi)$ is given by

$$Y_l^m(\theta, \phi) = \frac{(-1)^{l+m}}{2^l l!} \sqrt{\frac{(2l+1)}{4\pi} \frac{(l-m)!}{(l+m)!}} e^{im\phi} (\sin\theta)^m \frac{d^{l+m}}{d(\cos\theta)^{l+m}} (\sin\theta)^{2l}$$

12. Prove that the above representations imply that

$$[Y_l^m(\theta, \phi)]^* = (-1)^m Y_l^{-m}(\theta, \phi)$$

13. Express the spherical harmonics $Y_l^m(\theta, \phi)$ in terms of the associated Legendre polynomials[26] $P_l^m(\cos\theta)$:

$$Y_l^m(\theta, \phi) = \eta \sqrt{\frac{(2l+1)}{4\pi} \frac{(l-|m|)!}{(l+|m|)!}} e^{im\phi} P_l^m(\cos\theta)$$

where $\eta = (-1)^m$ for $m \geq 0$ and $\eta = 1$ when $m < 0$. The $P_l^m(x)$ are defined by

$$P_l^m(x) = (1-x^2)^{\frac{|m|}{2}} \left(\frac{d}{dx}\right)^{|m|} P_l(x)$$

$$\text{with} \qquad P_l(x) = \frac{(-1)^l}{2^l l!} \left(\frac{d}{dx}\right)^l (1-x^2)^l$$

where $P_l(x)$ are called the Legendre polynomials[27].

14. Work out the parity of the spherical harmonics: how does $Y_l^m(\theta, \phi)$ change under $\vec{r} \to -\vec{r}$?

(Hint: what does $\vec{r} \to -\vec{r}$ mean in terms of θ and ϕ?)

15. Consider a particle constrained to move on the surface of a sphere of fixed radius[28].

---

[26]This result can also be proved by first showing that $Y_l^0(\theta, \phi)$ can be expressed in terms of the Legendre polynomial $Y_l^0(\theta, \phi) = \sqrt{2l+1/4\pi} \, P_l(\cos\theta)$ and then using the fact that for $m \geq 0$, $Y_l^m(\theta, \phi)$ is proportional to $\left(\hat{L}_+\right)^m Y_l^0(\theta, \phi)$. Application of the formula in part (b) of problem 11 yields the final result.

[27]Note that, while for $m \geq 0$, one merely needs to plug in the value of $P_l^m(\cos\theta)$ to retrieve the formula for $Y_l^m(\theta, \phi)$ derived earlier, for negative m, one has to work a little more.

[28]To a first approximation, it is possible to treat the translational, vibrational and rotational degrees of freedom of a diatomic molecule independently. To treat the rotational part, the diatomic molecule is modelled as a rigid rotor: a system of two identical masses connected by a massless rigid rod. In the centre of mass frame this is equivalent to a particle (with mass equal to the reduced mass) moving on the surface of a sphere having radius equal to the inter-particle distance.

(a) Show that the stationary states of the system are the common eigenstates of $\hat{L}^2$ and $\hat{L}_z$.

(b) Determine the energy spectrum for this system and comment on the degeneracy.

(c) If the system is in the stationary state $|l = 2, m = 2\rangle$ (l and m being the orbital and azimuthal quantum numbers), find out the probability that the particle may be found in the positive octant.

Hydrogen Atom

As our final project, we will solve the eigenvalue problem for the Hamiltonian of the hydrogen atom. The classical picture of the hydrogen atom is a negatively charged particle (electron) bound to a positively charged particle (proton) by the coulomb potential. We shall actually solve for the Hamiltonian corresponding to the *internal energy* of the system. This is achieved by transforming to a new set of coordinates: the centre of mass $\vec{R}$ and relative coordinate of the electron with respect to the proton $\vec{r}$.

$$\vec{R} = \frac{m_p \vec{r}_p + m_e \vec{r}_e}{m_p + m_e} \qquad \text{and} \qquad \vec{r} = \vec{r}_e - \vec{r}_p$$

where m_p and $\vec{r}_p$ are the mass and position vector of the proton, and m_e and $\vec{r}_e$ are the corresponding quantities for the electron. The energy of the system, for a potential $V(r)$ that depends only on the relative coordinate, splits up into the centre of mass and relative coordinate parts:

$$H_{total} = \frac{1}{2M}P^2 + \frac{1}{2m}p^2 + V(r)$$

where $M = m_p + m_e$, $\vec{P} = M\left(d\vec{R}/dt\right)$, $m = (m_p m_e)/(m_p + m_e)$ and $\vec{p} = m\left(d\vec{r}/dt\right)$. Here, M is the total mass of the system, and m is called the reduced mass of the system. Also, $\vec{P}$ turns out to be the total momentum of the system, and $\vec{p}$ is called the relative momentum. Considering the internal energy H amounts to focussing on the relative coordinate part:

$$H = \frac{1}{2m}p^2 + V(r)$$

This is formally identical to a one body problem of a particle of mass m moving under the central potential $V(r)$. For the hydrogen atom, the potential is simply the coulomb potential $V(r) = -e^2/4\pi\epsilon_0 r$ produced by the positive charge $+e$ of the proton that attracts the electron having a negative charge $-e$. Here ϵ_0 is the permittivity of vaccum.

Now for a quantum particle moving under a central potential, the Hamiltonian operator

$$\hat{H} = \frac{1}{2m}\hat{p}^2 + V(\hat{r})$$

represented in position space using Cartesian coordinates is given by

$$\langle x, y, z| \frac{1}{2m}\hat{p}^2 + V(\hat{r}) |\psi\rangle = \left[-\frac{\hbar^2}{2m}\left(\frac{\partial^2}{\partial x^2} + \frac{\partial^2}{\partial y^2} + \frac{\partial^2}{\partial z^2} \right) + V(r) \right] \langle x, y, z|\psi\rangle$$

$$\left[\hat{H} \right]^{(x,y,z)} = -\frac{\hbar^2}{2m}\left(\frac{\partial^2}{\partial x^2} + \frac{\partial^2}{\partial y^2} + \frac{\partial^2}{\partial z^2} \right) + V(r)$$

Here we retained the symbol r in $V(r)$ which, obviously, stands for $\sqrt{x^2 + y^2 + z^2}$. We note that the position representation of $\hat{p}^2$ is essentially the Laplacian operator. Since the potential is central, and therefore spherically symmetric, it is convenient to go over to spherical polar coordinates (r, θ, ϕ). The representation of the Hamiltonian operator in the (r, θ, ϕ) space is

$$\left[\hat{H} \right]^{(r,\theta,\phi)} = -\frac{\hbar^2}{2m}\left(\frac{1}{r^2}\frac{\partial}{\partial r}\left(r^2\frac{\partial}{\partial r} \right) + \frac{1}{r^2 \sin\theta}\frac{\partial}{\partial\theta}\left(\sin\theta\frac{\partial}{\partial\theta} \right) + \frac{1}{r^2 \sin^2\theta}\frac{\partial^2}{\partial\phi^2} \right)$$
$$+ V(r)$$

The eigenvalue equation for the Hamiltonian, when cast in the (r, θ, ϕ) space, becomes

$$\left[\hat{H} \right]^{(r,\theta,\phi)} \langle r, \theta, \phi|E\rangle = E\langle r, \theta, \phi|E\rangle$$

Now, recall that the representation of the total orbital angular momentum squared operator in the (r, θ, ϕ) space is

$$\left[\hat{L}^2 \right]^{(r,\theta,\phi)} = -\frac{\hbar^2}{\sin^2\theta}\left[\sin\theta\frac{\partial}{\partial\theta}\left(\sin\theta\frac{\partial}{\partial\theta} \right) + \frac{\partial^2}{\partial\phi^2} \right]$$

Therefore, the Hamiltonian operator can be written as

$$\left[\hat{H} \right]^{(r,\theta,\phi)} = -\frac{\hbar^2}{2m}\left(\frac{1}{r^2}\frac{\partial}{\partial r}\left(r^2\frac{\partial}{\partial r} \right) - \frac{\left[\hat{L}^2 \right]^{(\theta,\phi)}}{\hbar^2 r^2} \right) + V(r)$$

Now, the operators involving only the radial coordinate r will commute with operators involving only the angular coordinates (θ, ϕ). Thus, it is evident that the Hamiltonian $\hat{H}$ will commute with $\hat{L}^2$ and $\hat{L}_z$, since $\hat{L}^2$ will commute with itself, and the $\hat{L}_z$ operator also commutes with $\hat{L}^2$. Hence, a complete set of common eigenvectors of $\left\{ \hat{H}, \hat{L}^2, \hat{L}_z \right\}$ will exist. With this observation let us try to solve the eigenvalue equation of the Hamiltonian using a trial solution of the form

$$\langle r, \theta, \phi|E\rangle \equiv \psi(r, \theta, \phi) = R(r)Y_l^m(\theta, \phi)$$

where $R(r)$ is a function of r alone, and $Y_l^m(\theta, \phi)$ are the spherical harmonics: the common eigenfunctions of $\left\{\hat{L}^2, \hat{L}_z\right\}$. Note that this is the most general possibility for a function to be a common eigenstate for $\left\{\hat{L}^2, \hat{L}_z\right\}$, since for a given l and m, the common eigenfunction $Y_l^m(\theta, \phi)$ is unique up to a multiplicative constant in the (θ, ϕ) space.

If we plug our trial solution back in the eigenvalue equation, we get

$$\left[-\frac{\hbar^2}{2m}\left(\frac{1}{r^2}\frac{d}{dr}\left(r^2\frac{d}{dr}\right) - \frac{l(l+1)}{r^2}\right) + V(r)\right] R_{E,l}(r) = E R_{E,l}(r)$$

where l is the azimuthal quantum number. Here we have also introduced the subscripts E and l (the parameters occurring in the differential equation) in $R_{E,l}(r)$. This is the radial equation for an arbitrary central potential[29].

For the specific case of the hydrogen atom, $V(r) = -Ze^2/4\pi\epsilon_0 r$, and the radial equation becomes

$$\left[-\frac{\hbar^2}{2m}\left(\frac{1}{r^2}\frac{d}{dr}\left(r^2\frac{d}{dr}\right) - \frac{l(l+1)}{r^2}\right) - \frac{Ze^2}{4\pi\epsilon_0 r}\right] R_{E,l}(r) = E R_{E,l}(r)$$

Here Z is the atomic number. Although $Z = 1$ for the hydrogen atom, it is customary to keep the Z in the formulae to facilitate easy generalization to *hydrogen like* atoms for which $Z \neq 1$.

If we make the substitutions

$$r^2 = -\rho^2\frac{\hbar^2}{8mE} \qquad \text{and} \qquad -\frac{Ze^2}{4\pi\epsilon_0}\frac{\sqrt{-8mE}}{4\hbar E} = n$$

then the radial equation can be mapped to the following standard differential equation:

$$\left[\frac{1}{\rho^2}\frac{d}{d\rho}\left(\rho^2\frac{d}{d\rho}\right) - \frac{l(l+1)}{\rho^2} + \frac{n}{\rho} - \frac{1}{4}\right] R_{n,l} = 0$$

that is usually discussed in connection with Laguerre differential equation; see, for example, (Margenau and Murphy 1956). This differential equation has well behaved, square integrable solutions only if n is a positive integer, and l is also integral. Moreover, for a given n, acceptable solutions exist only for $l < n$. Thus, the allowed values for n and l are

$$n = 1, 2, 3, \ldots \qquad \text{and} \qquad l = 0, 1, 2, \ldots, n - 1$$

The restriction on n gives the allowed values of the energy E (which we now write as E_n):

$$E_n = -\frac{m}{2\hbar^2}\left(\frac{Ze^2}{4\pi\epsilon_0}\right)^2\frac{1}{n^2}$$

---

[29]If we naively use separation of variables for the eigenvalue equation of the Hamiltonian, the angular equation will turn out to be simply the eigenvalue equation of $\hat{L}^2$. If we use separation of variables for this equation, we shall see that the ϕ equation will automatically be obeyed by the solution of the eigenvalue equation of $\hat{L}_z$.

For $n = 1$, we get the ground state energy E_1 of the hydrogen atom. The numerical value of $E_1 = -13.6 \ eV$. Incidentally, n is called the **principal quantum number**.

The degeneracy in E is easy to calculate. Since the eigenvalue of $\hat{L}^2$ associated with l is $2l + 1$ fold degenerate, the degeneracy g_n of E_n is[30]

$$g_n = \sum_{l=0}^{n-1} (2l + 1) = n^2$$

The solution $R_{n,l}$ to the radial equation, written in terms of ρ and replacing the subscript E by n, is

$$R_{n,l}(\rho) = N\rho^l e^{-\rho/2} L_{n-1-l}^{2l+1}(\rho)$$

where $L_{n-1-l}^{2l+1}(\rho)$ are known as **associated Laguerre polynomials**, and N is the constant of normalization[31]. The associated Laguerre polynomials L_{q-p}^p are defined by

$$L_{q-p}^p(\rho) = (-1)^p \left(\frac{d}{d\rho}\right)^p L_q(\rho)$$

where the $L_q(\rho)$ are the **Laguerre polynomials** given by

$$L_q(\rho) = e^\rho \left(\frac{d}{d\rho}\right)^q \left(e^{-\rho}\rho^q\right)$$

Finally, reverting to the radial coordinate r, the normalized radial function $R_{n,l}(r)$ has the form

$$R_{n,l}(r) = \sqrt{\left(\frac{2}{na}\right)^3 \frac{(n-l-1)!}{2n\left[(n+1)!\right]^3}} \left(\frac{2r}{na}\right)^l \exp\left\{-\frac{r}{na}\right\} L_{n-1-l}^{2l+1}\left(\frac{2r}{na}\right)$$

where a is a constant known as the *Bohr radius*[32]. The **Bohr radius** is given by $a = 4\pi\epsilon_0 \hbar^2/Ze^2 m$.

---

[30]Actually, we have ignored another observable called *spin* having purely quantum origin (which we have briefly encountered in the last problem set). Electrons can come with two possible spin values, and each electronic state can have two associated spin wave functions that will multiply its spatial wave function just as the angular wave functions Y_l^m multiply the radial wave functions $R_{E,l}$. For the hydrogen atom the energy does not depend on the spin (to a first approximation). This will throw in a factor of 2 to the degeneracy, so that the actual degeneracy is $2n^2$.

[31]It is customary to normalize the $R_{n,l}$ and Y_l^m separately.

[32]Bohr's theory was a remarkably successful attempt to describe the spectrum of the hydrogen atom through some ad hoc assumptions fused into an otherwise classical theory. According to this theory, the hydrogen atom was an electron revolving around the proton in circular orbits. The electron's orbital angular momentum could only take values which were integral multiples of $\hbar$. This implies that the allowed radii of the circular orbits were also discrete: given by $r_n = n^2 a_0$. Here n is a positive integer and a_0 is a constant known as the Bohr radius. Clearly, the Bohr radius is the radius of the orbit when the electron is in its lowest orbit $n = 1$.

The full wave function for the Hydrogen atom $\psi_{n,l,m}(r,\theta,\phi)$ is then the product of the radial function $R_{n,l}(r)$ defined above, and the spherical harmonics defined in the previous section[33]:

$$\psi_{n,l,m}(r,\theta,\phi) = R_{n,l}(r)\, Y_l^m(\theta,\phi)$$

Problems

1. A particle in a *spherical box* is defined by the potential:

$$V(r) = \begin{cases} 0 & r<a \\ \infty & r \geq a \end{cases}$$

   (a) Setup the eigenvalue equation for the Hamiltonian in position space and convince yourself that the formal solution to the equation can be written as

   $$\psi_{E,l,m}(r,\theta,\phi) = K_{E,l}(r)\, Y_l^m(\theta,\phi)$$

   where $K_{E,l}(r)$ denotes the radial part of the wave function and $Y_l^m(\theta,\phi)$ are the spherical harmonics[34]. Here E labels the different energies.

   (b) Simplify the radial equation by using the change of variables[35]:

   $$K_{E,l}(r) = \frac{u_{E,l}(r)}{r}$$

   Identify the equation for $l = 0$, and determine the corresponding wave functions and energies.

2. The ground state of the hydrogen atom is given by

$$R_{1,0} = 2a^{-\frac{3}{2}}\exp\left\{-\frac{r}{a}\right\} \qquad \text{and} \qquad Y_{0,0} = \frac{1}{\sqrt{4\pi}}$$

   where a is the Bohr radius.

---

[33] A detailed solution of the radial differential equation (employing the *power series* method) is worked out in many standard undergraduate text books. See, for example, (Griffiths 2005).

[34] Incidentally, solution to the radial part is given by the so called *spherical Bessel function* $J_l(kr)$, *i.e.*, $K_{E,l}(r) = J_l(kr)$ where $k = \sqrt{2mE}/\hbar$, m being the mass of the particle.

[35] This is quite a standard substitution employed for central potential problems. One, typically, writes the differential equation in terms of u, analyses the asymptotic behaviour, and factors out the dominant terms in the asymptotic limit. One then proceeds with a power series solution for the remaining factor and imposes the requirements of well-behavedness. Indeed, the treatment of the hydrogen atom follows this line in most books.

Show that in the ground state

(a) the most probable value of the radial distance is the Bohr radius a.

(b) the expectation value $\langle \hat{r} \rangle$ of the radial distance r is 1.5 times its most probable value a. You may use the standard integral:

$$\int_0^\infty x^n e^{-\alpha x}\, dx = \frac{n!}{\alpha^{n+1}}\,;\ \alpha > 0$$

(c) the expectation value of $\langle \hat{x} \rangle$ is zero (preferably without calculation).

3. Find out the expectation value of the kinetic energy $\hat{T}$ and the potential energy $\hat{V}$ in the ground state of the hydrogen atom. Can you think of a quick way to check your result?

4. The spherical harmonics, with an appropriate normalization, obey the condition

$$\sum_{m=-l}^{+l} |Y_l^m|^2 = \frac{2l+1}{4\pi}$$

What is the position probability density if a hydrogen atom is in a *statistical mixture* of all the $\psi_{n,l,m}$ states for $n \leq q$ (q is some positive integer) with all states being equally likely[36]. Do you expect it to be similar if the hydrogen atom is in an *unbiased* superposition of all the $\psi_{n,l,m}$ states (i.e., if the state is a superposition of all the possible $\psi_{n,l,m}$ states with $n \leq q$ where all the expansion coefficients are identical). Leave your answer in terms of the radial wave functions $R_{n,l}$.

5. Imagine that a hydrogen atom is in an unbiased superposition of states compatible with the conditions $n \leq 2$ and $l = 0$ at time $t = 0$.

(a) How will such a state evolve in time?

(b) Find out the nature of the time dependence of the expectation value of the radial distance r.

6. The hydrogen atom, being a system of two oppositely charged particles, should naturally have an electric dipole moment.

(a) Write down the electric dipole moment operator for the hydrogen atom.

---

[36]It is *not* to be confused with a superposition of the allowed states with coefficients having equal modulus. It means that the system is in any one of the allowed states, and the likelihood of each state is equal.

(b) Show that the expectation value of the dipole moment is zero in any stationary state.

(Hint: The parity of the hydrogenic wave functions $\psi_{n,l,m}$ is controlled by the relation $Y_l^m(-\vec{r}) = (-1)^l Y_l^m(\vec{r})$.)

(c) Find out the expectation value of the dipole moment if the initial state of the atom is $\psi|_{t=0} = (1/\sqrt{2})(\psi_{1,0,0} + \psi_{2,1,0})$. Use the relations:

$$x = \sqrt{\frac{2\pi}{3}}r\left(Y_1^{-1} - Y_1^1\right), \quad y = i\sqrt{\frac{2\pi}{3}}r\left(Y_1^{-1} + Y_1^1\right), \quad z = \sqrt{\frac{2\pi}{3}}rY_1^0$$

and the radial wave functions:

$$R_{1,0} = 2a^{-\frac{3}{2}}\exp\left\{-\frac{r}{a}\right\} \quad \text{and} \quad R_{2,1} = \frac{1}{\sqrt{24}}a^{-\frac{3}{2}}\frac{r}{a}\exp\left\{-\frac{r}{2a}\right\}$$

7. Although the full hydrogenic wave functions form fully orthogonal sets (i.e., with respect to all its indices), the radial wave functions are only partially orthogonal. Show this explicitly for $R_{1,0}$ and $R_{2,1}$.

8. A positronium is a hydrogen atom like system where an electron is bound to a *positron* instead of the proton. A positron is a particle which is identical to the electron except that its charge is positive. In particular it has the same mass as the electron.

(a) How would its spectrum compare with that of the hydrogen atom?

(b) Estimate the size of this system as compared to the hydrogen atom.

(Hint: Think about a reasonable measure to estimate the size of the atom. Recall that the mass m_e of the electron is much smaller than the mass m_p of the proton $(m_p = 1836\,m_e)$.)

Chapter 9

Composite Systems

From the very early days of science we have been asking, what is everything made up of? This question stems from the belief that matter as we know it, is made out of some *elementary building blocks*. Our theoretical paradigms try to conceive the vast diversity of phenomena that we see around us as a manifestation of complex combinations of some *elementary interactions between elementary constituents* of matter. In fact, when we ask "what is everything made up of?", the very question seeks an answer that must conform to this philosophical choice. It has dominated the entire course of scientific history. Thus, being able to deal with composite systems must be an essential virtue of any theory. In the preceding chapters we have outlined a formal description of QM. In this chapter we shall see how, within the framework that has been laid out, we can describe physical scenarios involving composite systems that is ubiquitous in physics.

The Meaning of a Composite System

Let us consider two quantum systems[1] S^1 and S^2. We will indicate the observables and states of the two systems using parenthesized superscripts '1' and '2' (e.g., $\hat{\alpha}^{(1)}, \hat{\beta}^{(2)}, |\psi^{(1)}\rangle, |\phi^{(2)}\rangle$, etc.). The entire machinery of QM applies individually to the two systems. Let the sets $\{\hat{\chi}^{(1)}\}$ and $\{\hat{\xi}^{(2)}\}$ comprise CSCO for the two systems[2]. Then we have, for example, definite probabilities $P\left[\chi_i^{(1)} \to \alpha_j^{(1)}\right]$, $P\left[\xi_k^{(2)} \to \beta_l^{(2)}\right]$, etc., defined on the systems. Now we can, quite trivially, imagine a *composite* system S simply by *juxtaposing* the two systems S^1 and S^2. Corresponding to the states $|\psi^{(1)}\rangle$ and $|\phi^{(2)}\rangle$ belonging to the systems S^1 and S^2, we can associate an ordered pair $(|\psi^{(1)}\rangle, |\phi^{(2)}\rangle)$, and

---

[1] For concreteness one could think of the two systems as two particles.
[2] The arguments that follow can easily be generalized to CSCO that have more than one member.

consider it to be a state of the composite system S. The *independent* events[3]

$$\chi_i^{(1)} \to \alpha_j^{(1)} \quad \text{and} \quad \xi_k^{(2)} \to \beta_l^{(2)}$$

for the systems S^1 and S^2 can be conceived as a single *elementary* event

$$\left(\chi_i^{(1)}, \xi_k^{(2)} \right) \to \left(\alpha_j^{(1)}, \beta_l^{(2)} \right)$$

for the composite system S. Hence, the probability of this event would be given by

$$P\left[\chi_i^{(1)}, \xi_k^{(2)} \to \alpha_j^{(1)}, \beta_l^{(2)} \right] = P\left[\chi_i^{(1)} \to \alpha_j^{(1)} \right] P\left[\xi_k^{(2)} \to \beta_l^{(2)} \right]$$

according to the rules of classical probability. It is now trivial to see that

$$P\left[\chi_i^{(1)} \to \alpha_j^{(1)} \right] = P\left[\chi_i^{(1)}, \xi_k^{(2)} \to \alpha_j^{(1)}, \xi_k^{(2)} \right]$$

Again, the probability of the event $\chi_i^{(1)} \to \alpha_j^{(1)}$ in S^1 can be written as

$$\begin{aligned}
P\left[\chi_i^{(1)} \to \alpha_j^{(1)} \right] &= P\left[\chi_i^{(1)} \to \alpha_j^{(1)} \right] \sum_l P\left[\xi_k^{(2)} \to \beta_l^{(2)} \right] \\
&= \sum_l P\left[\chi_i^{(1)} \to \alpha_j^{(1)} \right] P\left[\xi_k^{(2)} \to \beta_l^{(2)} \right] \\
&= \sum_l P\left[\chi_i^{(1)}, \xi_k^{(2)} \to \alpha_j^{(1)}, \beta_l^{(2)} \right] \\
&= P\left[\chi_i^{(1)}, \xi_k^{(2)} \to \alpha_j^{(1)} \right]
\end{aligned}$$

which is a probability for a *compound* event in the system S in which a measurement of $\hat{\alpha}^{(1)}$ yields $\alpha_j^{(1)}$ and $\hat{\beta}^{(2)}$ yields anything. Here, the observables $\hat{\xi}^{(2)}$ and $\hat{\beta}^{(2)}$ can be chosen arbitrarily and l runs over the entire spectrum of $\hat{\beta}^{(2)}$.

For our composite system, we thus have the following observations:

One can identify

- *with every observable in S^1 (or S^2), an observable in S,*

- *with every state in S^1 (or S^2), a set of states in S, and*

- *with every event in S^1 (or S^2) specific events in S such that the probabilities are identical.*

---

[3]If you are not sure of the *exact* meaning of an *independent event*, see appendix 'A'. There is more to it than what is intuitively suggested by the term *independent*.

One can also, in the same spirit, consider the time evolution of the states $\left|\psi^{(1)}\right\rangle$ and $\left|\phi^{(2)}\right\rangle$ belonging to the systems S^1 and S^2 respectively, to be associated with the time evolution of the ordered pair $\left(\left|\psi^{(1)}\right\rangle, \left|\phi^{(2)}\right\rangle\right)$, which is a state in the system S. If the states $\left|\psi^{(1)}(0)\right\rangle$ and $\left|\phi^{(2)}(0)\right\rangle$ evolve into the states $\left|\psi^{(1)}(t)\right\rangle$ and $\left|\phi^{(2)}(t)\right\rangle$ in time t, then we would be obliged to assume that the initial state $\left(\left|\psi^{(1)}(0)\right\rangle, \left|\phi^{(2)}(0)\right\rangle\right)$ of S evolves into the state $\left(\left|\psi^{(1)}(t)\right\rangle, \left|\phi^{(2)}(t)\right\rangle\right)$ at time t.

Now if we are given a quantum system (described by some inner product space) for which we are able to identify parts (i.e., sets of states and observables) such that the features and properties mentioned above hold, we could *want* to call it a composite system. However, it is obvious that a composite system defined merely by juxtaposing two systems is not expected to be useful or interesting. The parts of such a system are, so to speak, not *coupled* to each other.

It is only when a composite system exhibits some properties that are not derivable from the properties of its constituents, that the study of the composite system becomes worthwhile.

Thus, for a nontrivial composite system, in the first place, we do not want to restrict *all* states in the composite system to be ordered pairs. Furthermore, even if an initial state of the composite system is an ordered pair, we will not expect that the time evolution will be some sort of a simple *union of the independent time evolutions* in the constituent systems, yielding a final state that is an ordered pair of the time evolved initial states of the constituents. Finally, in the composite system we will not expect that *all* of the observables of the composite system will be related to one or the other of its parts. We will expect to see observables which are meaningful and definable only for the composite system *as a whole*.

Now we ask, whether it is possible to find a suitable mathematical framework that we can propose as a formal description of composite systems. Such a mathematical framework must, obviously, allow the description of trivially juxtaposed composite systems (whose parts are not coupled to each other) but must also be able to accommodate the nontrivial composite systems having features described in the preceding paragraph. We shall turn to study such a mathematical framework in the following section. One should, of course, always bear in mind that the validation of a proposal of a formal description must finally come from experiments.

Tensor Product Space

Tensor product of vectors and tensor product space

We consider two vector spaces V^1 and V^2. The set of all ordered pairs of vectors $\left(\left|\psi^{(1)}\right\rangle, \left|\phi^{(2)}\right\rangle\right)$ with $\left|\psi^{(1)}\right\rangle$ belonging to V^1 and $\left|\phi^{(2)}\right\rangle$ belonging to V^2 is called a ***Cartesian product***, and it is denoted by $V^1 \times V^2$. Now we define ***tensor product space*** $V^1 \otimes V^2$ as a vector space that has the following properties:

1. There exists a bilinear function[4] $\otimes : V^1 \times V^2 \rightarrow V^1 \otimes V^2$. This function is called a ***tensor product*** (to be distinguished from the tensor product space) and the value of the function is denoted by

$$\otimes \left(\left|\psi^{(1)}\right\rangle, \left|\phi^{(2)}\right\rangle\right) \equiv \left|\psi^{(1)}\right\rangle \otimes \left|\phi^{(2)}\right\rangle$$

Bilinearity of tensor product implies

   (a) $\left(c\left|\psi^{(1)}\right\rangle\right) \otimes \left|\phi^{(2)}\right\rangle = \left|\psi^{(1)}\right\rangle \otimes \left(c\left|\phi^{(2)}\right\rangle\right) = c\left(\left|\psi^{(1)}\right\rangle \otimes \left|\phi^{(2)}\right\rangle\right)$

   (b) $\left(\left|\psi_1^{(1)}\right\rangle + \left|\psi_2^{(1)}\right\rangle\right) \otimes \left|\phi^{(2)}\right\rangle = \left|\psi_1^{(1)}\right\rangle \otimes \left|\phi^{(2)}\right\rangle + \left|\psi_2^{(1)}\right\rangle \otimes \left|\phi^{(2)}\right\rangle$

   (c) $\left|\psi_1^{(1)}\right\rangle \otimes \left(\left|\phi_1^{(2)}\right\rangle + \left|\phi_2^{(2)}\right\rangle\right) = \left|\psi^{(1)}\right\rangle \otimes \left|\phi_1^{(2)}\right\rangle + \left|\psi^{(1)}\right\rangle \otimes \left|\phi_2^{(2)}\right\rangle$

Here, in the set of equations (a), $c \in \mathbb{C}$.

2. If $B^1 = \left\{\left|\chi_i^{(1)}\right\rangle ; i = 1, 2, \ldots, d_1\right\}$ and $B^2 = \left\{\left|\xi_j^{(2)}\right\rangle ; j = 1, 2, \ldots, d_2\right\}$ are bases in V^1 and V^2 (assumed to be of dimensions d_1 and d_2 respectively), then the set $B = \left\{\left|\chi_i^{(1)}\right\rangle \otimes \left|\xi_j^{(2)}\right\rangle ; i = 1, 2, \ldots, d_1 ; j = 1, 2, \ldots, d_2\right\}$ is a basis in[5] $V^1 \otimes V^2$.

For the tensor product $\left|\psi^{(1)}\right\rangle \otimes \left|\phi^{(2)}\right\rangle$ of two vectors $\left|\psi^{(1)}\right\rangle$ and $\left|\phi^{(2)}\right\rangle$, having the respective expansions

$$\left|\psi^{(1)}\right\rangle = \sum_{i=1}^{d_1} \psi^i \left|\chi_i^{(1)}\right\rangle \quad \text{and} \quad \left|\phi^{(2)}\right\rangle = \sum_{j=1}^{d_2} \phi^j \left|\xi_j^{(2)}\right\rangle$$

it is clear that

$$\left|\psi^{(1)}\right\rangle \otimes \left|\phi^{(2)}\right\rangle = \sum_{i=1}^{d_1} \sum_{j=1}^{d_2} \psi^i \phi^j \left(\left|\chi_i^{(1)}\right\rangle \otimes \left|\xi_j^{(2)}\right\rangle\right)$$

---

[4] A bilinear function is just a function of two variables that is linear in both.

[5] It can be shown that this definition is consistent: any choice of bases leads to the same tensor product space. Please see appendix 'B' for the proof.

However, it should be noted that not every vector in the tensor product space is a tensor product vector; a general vector in $V^1 \otimes V^2$ is some arbitrary linear combination of the basis vectors $\left|\chi_i^{(1)}\right\rangle \otimes \left|\xi_j^{(2)}\right\rangle$ in[6] B.

From here on, we shall often use a simpler notation $\left|\psi^{(1)}\right\rangle \left|\phi^{(2)}\right\rangle$, or sometimes even $\left|\psi^{(1)}, \phi^{(2)}\right\rangle$, to denote the tensor product $\left|\psi^{(1)}\right\rangle \otimes \left|\phi^{(2)}\right\rangle$.

Inner product on tensor product space

We define an ***inner product*** on the tensor product vectors in $V^1 \otimes V^2$ by

$$\left\langle \alpha^{(1)} \beta^{(2)} \middle| \psi^{(1)} \phi^{(2)} \right\rangle = \left\langle \alpha^{(1)} \middle| \psi^{(1)} \right\rangle \left\langle \beta^{(2)} \middle| \phi^{(2)} \right\rangle$$

This, obviously, specifies the inner products between all the basis elements of any basis in $V^1 \otimes V^2$ that is made out of tensor product vectors (e.g., the basis B used above in the definition of tensor product space). The inner product between arbitrary pairs of vectors in $V^1 \otimes V^2$ is then automatically specified by requiring the inner product to obey the defining properties of inner products, namely linearity and conjugate symmetry.

Linear operators on tensor product space

Now we turn to linear operators on the tensor product space $V^1 \otimes V^2$. If $\hat{\alpha}^{(1)}$ is an operator defined on V^1 and if $\hat{\beta}^{(2)}$ is an operator defined in V^2, then we define the ***tensor product operator*** $\hat{\alpha}^{(1)} \otimes \hat{\beta}^{(2)}$ by

$$\left(\hat{\alpha}^{(1)} \otimes \hat{\beta}^{(2)}\right) \left|\psi^{(1)}\right\rangle \left|\phi^{(2)}\right\rangle = \left(\hat{\alpha}^{(1)} \left|\psi^{(1)}\right\rangle\right) \otimes \left(\hat{\beta}^{(2)} \left|\phi^{(2)}\right\rangle\right)$$

Since the basis vectors of $V^1 \otimes V^2$ are tensor product vectors, it is clear that the requirement of linearity of the operators, together with the above condition ensures that the action of a tensor product operator on all vectors in $V^1 \otimes V^2$ are specified.

Once again, note that in $V^1 \otimes V^2$, just as all vectors are not tensor product vectors, all linear operators are also not tensor product operators.

Incidentally, the ***extension*** $\hat{\alpha}$ of an operator $\hat{\alpha}^{(1)}$ that acts on V^1 to the tensor product space V is defined by

$$\hat{\alpha} \left(\left|\psi^{(1)}\right\rangle \left|\phi^{(2)}\right\rangle\right) = \left(\hat{\alpha}^{(1)} \left|\psi^{(1)}\right\rangle\right) \otimes \left(\hat{I}^{(2)} \left|\phi^{(2)}\right\rangle\right)$$

where $\hat{I}^{(2)}$ is the identity operator acting on V^2. It is easy to check that the tensor product of two operators acting on V^1 and V^2 is equal to the ordinary product of their extensions to[7] $V^1 \otimes V^2$.

---

[6]In order to specify an arbitrary vector in $V^1 \otimes V^2$ we shall need to specify $d_1 \times d_2$ coefficients in general. If every vector were a tensor product, then the $d_1 \times d_2$ complex coefficients would always be describable in terms of $d_1 + d_2$ complex numbers!

[7]Please see appendix 'B' to check out some of the important properties of tensor product operators in regard to their eigenvalues and eigenvectors.

Description of Composite Systems

Now we are adequately equipped to put forward the assertion that embodies the description of composite systems:

- *The state space V of a composite system, constituted of systems S^1 and S^2, is the tensor product space $V^1 \otimes V^2$ where V^1 and V^2 are the state spaces of the respective component systems[8] S^1 and S^2.*

Let us now consider a composite system S made out of two component systems S^1 and S^2. If $\left\{\hat{\chi}^{(1)}\right\}$ and $\left\{\hat{\xi}^{(2)}\right\}$ comprise CSCO for these two systems, then a CSCO for the composite system S would be $\left\{\underline{\hat{\chi}}^{(1)}, \underline{\hat{\xi}}^{(2)}\right\}$ where $\underline{\hat{\chi}}^{(1)}$ and $\underline{\hat{\xi}}^{(2)}$ are the extensions of $\hat{\chi}^{(1)}$ and $\hat{\xi}^{(2)}$ to V. It is easy to see from the definition of extensions, that $\underline{\hat{\chi}}^{(1)}$ and $\underline{\hat{\xi}}^{(2)}$ must commute. Moreover, if $\left\{\left|\chi_i^{(1)}\right\rangle ; i = 1, 2, \ldots, d_1\right\}$ and $\left\{\left|\xi_j^{(2)}\right\rangle ; j = 1, 2, \ldots, d_2\right\}$ are bases constituted of orthonormal eigenvectors of $\hat{\chi}^{(1)}$ and $\hat{\xi}^{(2)}$ respectively, then the set $\left\{\left|\chi_i^{(1)}\right\rangle \left|\xi_j^{(2)}\right\rangle ; i = 1, 2, \ldots, d_1; j = 1, 2, \ldots, d_2\right\}$ will form an orthonormal basis in V. Hence we have

$$\underline{\hat{\chi}}^{(1)}\left(\left|\chi_i^{(1)}\right\rangle\left|\xi_j^{(2)}\right\rangle\right) = \left(\hat{\chi}^{(1)}\left|\chi_i^{(1)}\right\rangle\right) \otimes \left(\hat{I}^{(2)}\left|\xi_j^{(2)}\right\rangle\right)$$
$$= \chi_i \left(\left|\chi_i^{(1)}\right\rangle\left|\xi_j^{(2)}\right\rangle\right)$$

and in an exactly similar fashion

$$\underline{\hat{\xi}}^{(2)}\left(\left|\chi_i^{(1)}\right\rangle\left|\xi_j^{(2)}\right\rangle\right) = \left(\hat{I}^{(1)}\left|\chi_i^{(1)}\right\rangle\right) \otimes \left(\hat{\xi}^{(2)}\left|\xi_j^{(2)}\right\rangle\right)$$
$$= \xi_j \left(\left|\chi_i^{(1)}\right\rangle\left|\xi_j^{(2)}\right\rangle\right)$$

Thus $\left\{\underline{\hat{\chi}}^{(1)}, \underline{\hat{\xi}}^{(2)}\right\}$ has a complete set of common eigenvectors which constitutes an orthonormal basis in V, and the specification of the eigenvalues (χ_i, ξ_j) of $\hat{\chi}$ and $\hat{\xi}$ specifies a unique common eigenvector in that basis.

Now, the probability $P\left[\chi_i^{(1)}, \xi_k^{(2)} \to \alpha_j^{(1)}, \beta_l^{(2)}\right]$ will be given by

$$P\left[\chi_i^{(1)}, \xi_k^{(2)} \to \alpha_j^{(1)}, \beta_l^{(2)}\right] = \left|\left\langle \alpha_j^{(1)} \beta_l^{(2)} | \chi_i^{(1)} \xi_k^{(2)} \right\rangle\right|^2$$
$$= \left|\left\langle \alpha_j^{(1)} | \chi_i^{(1)} \right\rangle\right|^2 \left|\left\langle \beta_l^{(2)} | \xi_k^{(2)} \right\rangle\right|^2$$
$$= P\left[\chi_i^{(1)} \to \alpha_j^{(1)}\right] P\left[\xi_k^{(2)} \to \beta_l^{(2)}\right]$$

---

[8]The rule is naturally generalized to the case of composite systems constituted of more than two components.

as required. Again, the probability $P\left[\chi_i^{(1)}, \xi_k^{(2)} \to \alpha_j^{(1)}\right]$ for getting the outcome α_j upon measurement of $\hat{\alpha}^{(1)}$ on the tensor product state $\left|\chi_i^{(1)}, \xi_k^{(2)}\right\rangle$ is given by

$$
\begin{aligned}
P\left[\chi_i^{(1)}, \xi_k^{(2)} \to \alpha_j^{(1)}\right] &= \sum_l \left|\left\langle \alpha_j^{(1)} \beta_l^{(2)} | \chi_i^{(1)} \xi_k^{(2)} \right\rangle\right|^2 \\
&= \left|\left\langle \alpha_j^{(1)} | \chi_i^{(1)} \right\rangle\right|^2 \sum_l \left|\left\langle \beta_l^{(2)} | \xi_k^{(2)} \right\rangle\right|^2 \\
&= P\left[\chi_i^{(1)} \to \alpha_j^{(1)}\right]
\end{aligned}
$$

as required. Here, l runs over the entire spectrum of $\hat{\beta}^{(2)}$ and we have also assumed (for simplicity) that $\left\{\hat{\alpha}^{(1)}\right\}$ and $\left\{\hat{\beta}^{(2)}\right\}$ comprise CSCO for the two systems S^1 and S^2.

So far, the exercise was only to show that the required properties of a composite system, as laid down in the first section of this chapter, is indeed, reproduced by our proposed mathematical characterization of a composite system. But as we have already remarked, the artificial sewing of two systems to make it look like a composite does not buy us anything new. The above properties are aspects of *trivial* composite systems, and they are only necessary conditions that a proposed description of composite system must obey. We shall have to ask, what are the new, *nontrivial* consequences implied by our formal definition. We shall now point out the features of our description of a composite system that *cannot* be constructed out of the properties of its constituents. These features, that we list below, have all been observed in real life composite systems.

1. We have seen that there can exist linear operators acting on V which are not tensor products of operators acting on V^1 and V^2. This means that there can be observables that are associated with the composite system which *cannot be split up into parts* that are associated with its constituents.

2. We have also noted that there can exist vectors (and therefore composite system states) in $V = V^1 \otimes V^2$ which are not tensor products of vectors in V^1 and V^2. These states which are not tensor products are called **entangled states**. We shall see that when a composite system inhabits an entangled state, the subsystems become *inextricably correlated* with each other. Such correlations imply one of the most stark and counterintuitive aspects of QM which we shall discuss below. The properties of an entangled state of a composite system can no longer be deduced from the states of the constituents.

3. We shall see that our description allows for a time evolution involving *interactions* (between subsystems), that is *nontrivial*. In particular, we

shall see that interactions can cause an unentangled state to evolve into an entangled state. In fact, it is *only* if interactions are involved, that an entangled state can be produced.

To demonstrate the last point, we have to first understand what we mean by "interaction" between parts of a system. This is what we shall address in the next section.

Interaction between Subsystems

It is actually more convenient to start by defining what we mean by *non-interacting systems*. Let us, once again, consider a composite system S made up of two systems S^1 and S^2. The state space V of S is a tensor product of the state spaces of S^1 and S^2 denoted by V^1 and V^2 respectively. Let the (time independent) Hamiltonians of the two subsystems be $\hat{H}^{(1)}$ and $\hat{H}^{(2)}$, so that the time evolution for the two systems are given by

$$\left|\psi^{(1)}(t)\right\rangle = e^{-\frac{i}{\hbar}\hat{H}^{(1)}t}\left|\psi^{(1)}(0)\right\rangle$$

$$\left|\psi^{(2)}(t)\right\rangle = e^{-\frac{i}{\hbar}\hat{H}^{(2)}t}\left|\psi^{(2)}(0)\right\rangle$$

Now, intuitively, if we want the systems to be non-interacting, it should make no difference whether we consider the systems S^1 and S^2 separately or as a composite. In order to ensure that all observable effects remains the same irrespective of this choice, we will have to require that a tensor product state $\left|\psi^{(1)}(0)\right\rangle\left|\psi^{(2)}(0)\right\rangle$ evolves into the tensor product state $\left|\psi^{(1)}(t)\right\rangle\left|\psi^{(2)}(t)\right\rangle$. It is easy to see that if we use the tensor product of the evolution operators in V^1 and V^2 as the evolution operator in V, the time evolution of any tensor product state in V will have this desired property:

$$\left(e^{-\frac{i}{\hbar}\hat{H}^{(1)}t}\otimes e^{-\frac{i}{\hbar}\hat{H}^{(2)}t}\right)\left|\psi^{(1)}(0)\right\rangle\left|\psi^{(2)}(0)\right\rangle$$

$$= \left(e^{-\frac{i}{\hbar}\hat{H}^{(1)}t}\left|\psi^{(1)}(0)\right\rangle\right)\left(e^{-\frac{i}{\hbar}\hat{H}^{(2)}t}\left|\psi^{(2)}(0)\right\rangle\right)$$

$$= \left|\psi^{(1)}(t)\right\rangle\left|\psi^{(2)}(t)\right\rangle$$

Now, we have mentioned earlier that the tensor product of two operators is equal to the ordinary product of the extensions of the two operators, so

$$e^{-\frac{i}{\hbar}\hat{H}^{(1)}t}\otimes e^{-\frac{i}{\hbar}\hat{H}^{(2)}t} = e^{-\frac{i}{\hbar}\underline{\hat{H}}^{(1)}t}e^{-\frac{i}{\hbar}\underline{\hat{H}}^{(2)}t}$$

where $\underline{\hat{H}}^{(1)}$ and $\underline{\hat{H}}^{(2)}$ are the extensions of the Hamiltonians $\hat{H}^{(1)}$ and $\hat{H}^{(2)}$ to V. We have also mentioned before that extensions of operators acting on two different spaces will commute. This allows us to write[9]

$$e^{-\frac{i}{\hbar}\underline{\hat{H}}^{(1)}t}e^{-\frac{i}{\hbar}\underline{\hat{H}}^{(2)}t} \;=\; e^{-\frac{i}{\hbar}\left(\underline{\hat{H}}^{(1)}+\underline{\hat{H}}^{(2)}\right)t}$$

Thus, the Hamiltonian in V will be

$$\hat{H} \;=\; \underline{\hat{H}}^{(1)} + \underline{\hat{H}}^{(2)}$$

In general, if this scenario prevails we will say that the systems S^1 and S^2 are *non-interacting*. Thus,

*two systems S^1 and S^2 will be called **non-interacting** if their composite S has a time evolution governed by a Hamiltonian $\hat{H}$ given by*

$$\hat{H} \;=\; \underline{\hat{H}}^{(1)} + \underline{\hat{H}}^{(2)}$$

where $\underline{\hat{H}}^{(1)}$ and $\underline{\hat{H}}^{(2)}$ are the extensions of the Hamiltonians $\hat{H}^{(1)}$ and $\hat{H}^{(2)}$ of the component systems S^1 and S^2 to the state space of[10] S.

Having defined non-interacting systems, we can now move to define *"interaction"* between systems as follows:

Interaction *An interaction of two systems is that component of the theory which causes the time evolution of their composite to be different from what it would have been if the systems were non-interacting.*

In the present context, it should be evident that if we add to the Hamiltonian $\hat{H}$ a term $\hat{H}'$, so that

$$\hat{H} = \underline{\hat{H}}^{(1)} + \underline{\hat{H}}^{(2)} + \hat{H}'$$

then a tensor product state $\left|\psi^{(1)}(0)\right\rangle \left|\psi^{(2)}(0)\right\rangle$ in V will not evolve into the state $\left|\psi^{(1)}(t)\right\rangle \left|\psi^{(2)}(t)\right\rangle$. Thus the term $\hat{H}'$ in the above equation, which

---

[9]This follows from the Baker-Campbell-Hausdorff formula: For two operators $\hat{A}$ and $\hat{B}$

$$e^{\hat{A}}e^{\hat{B}} = \exp\left\{\hat{A} + \hat{B} + \frac{1}{2}\left[\hat{A},\hat{B}\right] + \frac{1}{12}\left(\left[\hat{A},\left[\hat{A},\hat{B}\right]\right] + \left[\hat{B},\left[\hat{B},\hat{A}\right]\right]\right) + \ldots\right\}$$

The '…' denotes terms involving nested commutators all of which have $\left[\hat{A},\hat{B}\right]$ at the deepest level (as in the third term).

[10]Here we have assumed that the systems involved are all conservative. The criterion can easily be generalized for time dependent Hamiltonians. We can simply proceed by changing the condition to require that the time evolution operator $\hat{U}$ in V can be written as a tensor product $\hat{U} = \hat{U}^{(1)} \otimes \hat{U}^{(2)}$ where $\hat{U}^{(1)}$ and $\hat{U}^{(2)}$ are the time evolution operators in V^1 and V^2.

is given by

$$\hat{H}' = \hat{H} - \left(\underline{\hat{H}}^{(1)} + \underline{\hat{H}}^{(2)} \right)$$

is defined to be the *interaction* for all practical purposes[11].

The above definition of "interaction" trivially leads to the following result.

- *If we start from a tensor product state, then the state can evolve into an entangled state only if there is interaction.*

Measurement on a Subsystem

Now that we have mathematically specified what we mean by composite system, we can talk about the rules that will enable us to answer the target questions of QM in the context of composite systems. The basic postulates of QM remain the same. The only additional component is that, now we have agreed to associate with a composite system, the tensor product space of the inner product spaces that describe the component subsystems.

Final state after a measurement on one part

Noting that the basic postulate that determines the final state after a measurement remains the same, let us apply it to a composite system S made out of two component systems S^1 and S^2. The state of the system soon after the measurement on an initial state $|\psi\rangle$, of an observable $\hat{\alpha}^{(1)}$ associated with the first subsystem, yielding an outcome $\alpha_{i'}$ would be given by

$$|\psi\,;\,\alpha_{i'}\rangle \;\; = \;\; \frac{\sum_{j,k} |\alpha_{i'}, \beta_j, \gamma_k\rangle \langle \alpha_{i'}, \beta_j, \gamma_k | \psi \rangle}{\sqrt{\sum_{j,k} |\langle \alpha_{i'}, \beta_j, \gamma_k | \psi \rangle|^2}}$$

where we have assumed that $\left\{ \hat{\alpha}^{(1)}, \hat{\beta}^{(1)} \right\}$ forms a CSCO in S^1, and $\left\{ \hat{\gamma}^{(2)} \right\}$ forms a CSCO in S^2, so that $\left\{ \hat{\alpha}^{(1)}, \hat{\beta}^{(1)}, \hat{\gamma}^{(2)} \right\}$ forms a CSCO in S (where the operators are now understood to be extensions to S). Here $|\alpha_{i'}, \beta_j, \gamma_k\rangle = \left| \alpha_{i'}^{(1)} \right\rangle \otimes \left| \beta_j^{(1)} \right\rangle \otimes \left| \gamma_k^{(2)} \right\rangle$, and the indices j and k obviously run over all the

---

[11]It may have occurred to the reader that, if we add to the non-interacting Hamiltonian $\underline{\hat{H}}^{(1)} + \underline{\hat{H}}^{(2)}$ a term such as $\underline{\hat{A}}^{(1)}$, that is an extension of some observable $\hat{A}^{(1)}$ of S^1, for example, then the time evolution will still be different from what it would be for the non-interacting situation, although an initial tensor product state will, in this case, evolve into some tensor product state. Should a term such as $\underline{\hat{A}}^{(1)}$ be regarded as an interaction? Theoretically, such a scenario may be imagined to be the result of the *composition* of a new system S^1_{new} with S^2, where S^1_{new} has a Hamiltonian $\hat{H}^{(1)} + \hat{A}^{(1)}$.

allowed values of the spectra of the respective observables, as usual. We can straight away make the following two observations:

1. If the initial state is a tensor product $|\psi\rangle = |\theta^{(1)}\rangle \otimes |\phi^{(2)}\rangle$ then so is the final state:

$$|\psi\,;\alpha_{i'}\rangle = \left(\frac{\sum_j |\alpha_{i'},\beta_j\rangle\,\langle\alpha_{i'},\beta_j|\theta^{(1)}\rangle}{\sqrt{\sum_j |\langle\alpha_{i'},\beta_j|\psi\rangle|^2}}\right) \otimes |\phi^{(2)}\rangle$$

The state of the second system $|\phi^{(2)}\rangle$ is simply unaffected.

2. Even when the initial state is not a tensor product state, the final state would still be a tensor product state *if a complete set of measurements is performed on one of the subsystems.* So if a measurement of $\hat{\alpha}^{(1)}$ and $\hat{\beta}^{(1)}$ on S^1 yields the outcomes $\alpha_{i'}$ and $\beta_{j'}$, we have

$$|\psi\,;\alpha_{i'},\beta_{j'}\rangle = |\alpha_{i'},\beta_{j'}\rangle \otimes \left(\frac{\sum_k |\gamma_k\rangle\,\langle\alpha_{i'},\beta_{j'},\gamma_k|\psi\rangle}{\sqrt{\sum_k |\langle\alpha_{i'},\beta_{j'},\gamma_k|\psi\rangle|^2}}\right)$$

Note that in the final tensor product state, the factor that corresponds to the system S^2 (i.e., the part in the parenthesis in the above expression) crucially depends on the outcome of the measurement on the first part.

This is a remarkable result because it holds even for subsystems that have arbitrary spatial separation. So, for systems that have interacted in the past (to produce an entangled state) and then have drifted away to far off places, if we perform a measurement on one part, that will *instantaneously* affect the state of the other part[12].

Probability in measurements on one part

We have already seen that, if on a tensor product state, we make a measurement of an observable that relates to one part of a system, then the probability of outcomes are independent of the other part of the system in every way. The probability is simply equal to what it would have been if the measurement was made on one part of the system forgetting completely about the other.

When the state of the system is entangled (i.e., not expressible as a tensor product) the probability of results on a part are not independent of the other parts. So one says that the systems are correlated. *When two systems*

---

[12]This *apparent* contradiction with special relativity (that allows news to travel only as fast as light) was pointed out by Einstein, Podolsky and Rosen in a famous 1935 paper. It is referred to in the literature as the EPR paradox. The phenomena has now been adequately validated by experiment. It has also been appropriately interpreted to show that there is really no contradiction with special relativity. Today, consequences such as the EPR paradox are no longer considered to be a *problem* of quantum mechanics but only a *feature*.

are correlated, the probability of outcomes of measurement on one part of the system will, in general, depend on the other part[13]. This is easy to see from the last formula of the previous subsection[14]. If $\hat{a}^{(1)}$ alone forms a CSCO in S^1, then the probability $P\left[|\psi\rangle \rightarrow \alpha_{i'}\right]$ would be given by $|\langle\psi\,;\alpha_{i'}|\psi\rangle|^2$ which depends on the amplitudes $\langle\alpha_{i'},\gamma_k|\psi\rangle$ which involves the states $\left|\gamma_k^{(2)}\right\rangle$.

We wish to conclude this section by mentioning that the features that we have discussed above have been overwhelmingly validated by experiments.

Examples of Composite Systems

A particle living in 2-dimensions

A particle living in a 1-dimensional space was described using the Hilbert space V^x having a basis $\{|x\rangle\,;\,x \in (-\infty,+\infty)\}$. We can similarly imagine a space V^y having basis $\{|y\rangle\,;\,y \in (-\infty,+\infty)\}$ corresponding to the coordinate y. A tensor product space $V^x \otimes V^y$ can now be constructed with a basis $\{|x\rangle\,|y\rangle\,;\,x,y \in (-\infty,+\infty)\}$. Now, it is easy to see that the space $V^{x,y}$ having a basis $\{|x,y\rangle\,;\,x,y \in (-\infty,+\infty)\}$ that was used to describe a particle moving in two dimensions can be identified with the tensor product space $V^x \otimes V^y$ and the basis in $V^{x,y}$ can be identified with the basis in $V^x \otimes V^y$.

$$V^{x,y} = V^x \otimes V^y \qquad \text{and} \qquad |x,y\rangle = |x\rangle\,|y\rangle$$

The basis vectors $|x\rangle\,|y\rangle$ are clearly common eigenvectors of $\hat{x}$ and $\hat{y}$ extended to $V^x \otimes V^y$. A general vector in $V^x \otimes V^y$ can be expanded in this basis formally as

$$|\psi\rangle = \int_{-\infty}^{+\infty} \int_{-\infty}^{+\infty} dx\,dy\, \psi\,(x,y)\,|x\rangle\,|y\rangle$$

where the function $\psi\,(x,y)$ is the representation of the vector in the basis. Vectors whose representations are square integrable functions qualify as quantum states, as usual. Since in the spaces V^x and V^y

$$\langle x'|x\rangle = \delta\,(x - x') \qquad \text{and} \qquad \langle y'|y\rangle = \delta\,(y - y')$$

the inner product in $V^x \otimes V^y$ will obey

$$\langle x',y'|x,y\rangle = \langle x'|x\rangle\,\langle y'|y\rangle = \delta\,(x - x')\,\delta\,(y - y')$$

---

[13]See problem 1(f).

[14]Incidentally, there is actually a formalism that uses linear operators instead of normalized vectors to describe quantum states. These operators are called *density operators*. In this formalism it is possible to assign density operators (by an operation called *partial trace*) to individual parts even if the system is in an entangled state. However, it is still *impossible* to describe the time evolution of the entangled state (i.e., the density operator of the composite state) in terms of the time evolution of the density operators of the individual parts.

So, the basis $\{|x\rangle\,|y\rangle\,;\,x,y \in (-\infty,+\infty)\}$ is orthonormal (in the extended sense) as before.

The position observables (extended to $V^x \otimes V^y$) are again represented as multiplicative operators

$$[\hat{x}]^{x,y} = x \quad ; \quad [\hat{y}]^{x,y} = y$$

and the momenta (extended to $V^x \otimes V^y$) are represented as the differential operators

$$[\hat{p}_x]^{x,y} = -i\hbar\frac{\partial}{\partial x} \quad ; \quad [\hat{p}_y]^{x,y} = -i\hbar\frac{\partial}{\partial y}$$

Finally, every observable $\hat{\theta}$ corresponds to some classical dynamical variable

$$\theta\left(x, p_x, y, p_y\right)$$

and is given by the same function with the coordinates and momenta replaced by the corresponding operators

$$\hat{\theta} = \theta\left(\hat{x}, \hat{p}_x, \hat{y}, \hat{p}_y\right)$$

Thus, we see that an appropriate way to describe a particle moving in two dimensions is to use a tensor product space of two spaces corresponding to each of the dimensions. The above discussion and results can be trivially generalized to higher dimensions.

Multiparticle systems

It is straightforward to quantize systems that are constituted of more than one particle. Suppose we have a two particle system (moving in 3 dimensions). Classically, the system will be described by six coordinates $\left(x^1, y^1, z^1, x^2, y^2, z^2\right)$ and six momenta $\left(p_x^1, p_y^1, p_z^1, p_x^2, p_y^2, p_z^2\right)$, where the superscripts indicate whether a variable refers to the first or the second particle. A CSCO for this system could be the set of six position observables $\left\{\hat{x}^1, \hat{y}^1, \hat{z}^1, \hat{x}^2, \hat{y}^2, \hat{z}^2\right\}$. We can assume, for example, the spectrum of each observable to be the real interval $(-\infty, +\infty)$. An appropriate space for the description of this system is the tensor product space $V^{x^1} \otimes V^{y^1} \otimes V^{z^1} \otimes V^{x^2} \otimes V^{y^2} \otimes V^{z^2}$. We shall often denote the coordinates of the i-th particle collectively by $\vec{r}^{\,i}$ where i can take the values 1 or 2. Hence, we shall write the above tensor product space as $V^{\vec{r}^{\,1}, \vec{r}^{\,2}} = V^{\vec{r}^{\,1}} \otimes V^{\vec{r}^{\,2}}$ where $V^{\vec{r}^{\,i}} \equiv V^{x^i} \otimes V^{y^i} \otimes V^{z^i}$. The basis vectors in $V^{\vec{r}^{\,1}, \vec{r}^{\,2}}$ are $\left|\vec{r}^{\,1}, \vec{r}^{\,2}\right\rangle = \left|\vec{r}^{\,1}\right\rangle\left|\vec{r}^{\,2}\right\rangle$ where $\left|\vec{r}^{\,i}\right\rangle = \left|x^i\right\rangle\left|y^i\right\rangle\left|z^i\right\rangle$. An arbitrary vector in this space is expanded as

$$|\psi\rangle = \int_{\vec{r}^{\,1}} \int_{\vec{r}^{\,2}} d^3\vec{r}^{\,1} d^3\vec{r}^{\,2}\, \psi\left(\vec{r}^{\,1}, \vec{r}^{\,2}\right) \left|\vec{r}^{\,1}, \vec{r}^{\,2}\right\rangle$$

In the $(\vec{r}^{\,1}, \vec{r}^{\,2})$ space this vector is represented as a function of six variables. Square integrable functions will describe quantum states as usual. The orthonormality condition becomes

$$\left\langle \vec{r}^{\,1'}, \vec{r}^{\,2'} \middle| \vec{r}^{\,1}, \vec{r}^{\,2} \right\rangle = \delta^3\left(\vec{r}^{\,1} - \vec{r}^{\,1'}\right) \delta^3\left(\vec{r}^{\,2} - \vec{r}^{\,2'}\right)$$

$$\text{where} \quad \delta^3\left(\vec{r}^{\,i} - \vec{r}^{\,i'}\right) = \delta\left(x^i - x^{i'}\right) \delta\left(y^i - y^{i'}\right) \delta\left(z^i - z^{i'}\right)$$

The position observables (extended to $V^{\vec{r}^{\,1}} \otimes V^{\vec{r}^{\,2}}$) are represented as multiplicative operators

$$\left[\hat{x}^i\right]^{\vec{r}^{\,1}, \vec{r}^{\,2}} = x^i \quad ; \quad \left[\hat{y}^i\right]^{\vec{r}^{\,1}, \vec{r}^{\,2}} = y^i \quad ; \quad \left[\hat{z}^i\right]^{\vec{r}^{\,1}, \vec{r}^{\,2}} = z^i$$

The momenta (extended to $V^{\vec{r}^{\,1}} \otimes V^{\vec{r}^{\,2}}$) are represented as differential operators

$$\left[\hat{p}^i_x\right]^{\vec{r}^{\,1}, \vec{r}^{\,2}} = -i\hbar \frac{\partial}{\partial x^i} \quad ; \quad \left[\hat{p}^i_y\right]^{\vec{r}^{\,1}, \vec{r}^{\,2}} = -i\hbar \frac{\partial}{\partial y^i} \quad ; \quad \left[\hat{p}^i_z\right]^{\vec{r}^{\,1}, \vec{r}^{\,2}} = -i\hbar \frac{\partial}{\partial z^i}$$

Finally, every observable $\hat{\theta}$ corresponds to some classical dynamical variable

$$\theta\left(x^1, p^1_x, y^1, p^1_y, z^1, p^1_z, x^2, p^2_x, y^2, p^2_y, z^2, p^2_z\right)$$

and is given by the same function with the coordinates and momenta replaced by the corresponding operators

$$\hat{\theta} = \theta\left(\hat{x}^1, \hat{p}^1_x, \hat{y}^1, \hat{p}^1_y, \hat{z}^1, \hat{p}^1_z, \hat{x}^2, \hat{p}^2_x, \hat{y}^2, \hat{p}^2_y, \hat{z}^2, \hat{p}^2_z\right)$$

In the following section we will have more to say about multiparticle systems.

Problems

1. Consider two, 2-level systems, S^1 and S^2, whose states live in similar inner product spaces V^1 and V^2 respectively. There is a "colour" observable for both the systems having identical spectrum $\{\text{blue}, \text{green}\} \equiv \{b, g\}$ so that we have orthonormal bases $C^1 = \{|b^1\rangle, |g^1\rangle\}$ and $C^2 = \{|b^2\rangle, |g^2\rangle\}$ in V^1 and V^2. We can imagine a composite system S whose states live in the tensor product space $V = V^1 \otimes V^2$ having a basis[15]

$$C = \left\{|b^1\rangle |b^2\rangle, |b^1\rangle |g^2\rangle, |g^1\rangle |b^2\rangle, |g^1\rangle |g^2\rangle\right\}$$

---

[15]In this problem, we have slightly deviated from the notation used in the text. For example, to avoid clutter, we have dispensed with the parenthesis in the superscripts. Consequently we have had to demote the component index of vectors downstairs as subscripts, breaking the convention used in the rest of the book.

We choose to denote the basis states as follows

$$
\begin{aligned}
|b^1\rangle\,|b^2\rangle &\equiv |bb\rangle \\
|b^1\rangle\,|g^2\rangle &\equiv |bg\rangle \\
|g^1\rangle\,|b^2\rangle &\equiv |gb\rangle \\
|g^1\rangle\,|g^2\rangle &\equiv |gg\rangle
\end{aligned}
$$

(a) Show that in order that a vector

$$
|\psi\rangle = p\,|bb\rangle + q\,|bg\rangle + r\,|gb\rangle + s\,|gg\rangle
$$

is a tensor product, the expansion coefficients p, q, r and s must satisfy the condition

$$
\frac{p}{q} = \frac{r}{s}
$$

(b) Show that if the vectors $|\chi^1\rangle$ and $|\xi^2\rangle$ have the representations

$$
\left[\chi^1\right]^{C^1} = \begin{bmatrix} \chi_1^1 \\ \chi_2^1 \end{bmatrix}
\qquad \text{and} \qquad
\left[\xi^2\right]^{C^2} = \begin{bmatrix} \xi_1^2 \\ \xi_2^2 \end{bmatrix}
$$

in the C^1 and C^2 spaces, the tensor product vector $|\chi^1\rangle\,|\xi^2\rangle$ will be represented as

$$
\left[\chi^1\xi^2\right]^{C} =
\begin{bmatrix}
\chi_1^1 \left(\begin{array}{c} \xi_1^2 \\ \xi_2^2 \end{array} \right) \\[12pt]
\chi_2^1 \left(\begin{array}{c} \xi_1^2 \\ \xi_2^2 \end{array} \right)
\end{bmatrix}
$$

in the C space.

(c) Show that if two operators $\hat{A}^1$ and $\hat{B}^2$ acting on V^1 and V^2 are represented as

$$
\left[\hat{A}^1\right]^{C^1} = \begin{bmatrix} A_{11}^1 & A_{12}^1 \\ A_{21}^1 & A_{22}^1 \end{bmatrix}
\qquad \text{and} \qquad
\left[\hat{B}^2\right]^{C^2} = \begin{bmatrix} B_{11}^2 & B_{12}^2 \\ B_{21}^2 & B_{22}^2 \end{bmatrix}
$$

in the bases C^1 and C^2 respectively, then the representation of the tensor product operator $\hat{A}^1 \otimes \hat{B}^2$ in the C basis is

$$
\left[\hat{A}^1 \otimes \hat{B}^2\right]^{C} =
\begin{bmatrix}
A_{11}^1 \left(\begin{array}{cc} B_{11}^2 & B_{12}^2 \\ B_{21}^2 & B_{22}^2 \end{array} \right) &
A_{12}^1 \left(\begin{array}{cc} B_{11}^2 & B_{12}^2 \\ B_{21}^2 & B_{22}^2 \end{array} \right) \\[18pt]
A_{21}^1 \left(\begin{array}{cc} B_{11}^2 & B_{12}^2 \\ B_{21}^2 & B_{22}^2 \end{array} \right) &
A_{22}^1 \left(\begin{array}{cc} B_{11}^2 & B_{12}^2 \\ B_{21}^2 & B_{22}^2 \end{array} \right)
\end{bmatrix}
$$

(d) If we label the outcomes of the colour observable so that $b = +\kappa$ and $g = -\kappa$ where κ is a real number, then in the spaces V^1 and V^2, the colour observables $\hat{C}^i$ $(i = 1, 2)$ obey

$$\hat{C}^i \left| b^i \right\rangle = +\kappa \left| b^i \right\rangle$$
$$\hat{C}^i \left| g^i \right\rangle = -\kappa \left| g^i \right\rangle$$

Write down the representations of the extensions $\underline{\hat{C}}^i$ of $\hat{C}^i$ to V in the basis C.

(e) If an observable $\hat{F}^1$ acting on V^1 is defined by

$$\hat{F}^1 \left| b^1 \right\rangle = +\kappa \left| g^1 \right\rangle$$
$$\hat{F}^1 \left| g^1 \right\rangle = +\kappa \left| b^1 \right\rangle$$

Find the representation of the extension $\underline{\hat{F}}^1$ of $\hat{F}^1$ to V in the basis C. Determine the spectrum of $\underline{\hat{F}}^1$, the degeneracy of the eigenvalues and identify a maximal set of independent eigenvectors for each of the eigenvalues.

(f) Suppose that a measurement of colour $\hat{C}^1$ associated with the system S^1 is performed that leads to an outcome blue b^1. What would be the final state and the probability of this event if the initial state is

   i. a tensor product: $\left| \psi \right\rangle = \left| \theta^1 \right\rangle \otimes \left| \phi^2 \right\rangle$?
   ii. a general state: $\left| \psi \right\rangle = \psi_{bb} \left| bb \right\rangle + \psi_{bg} \left| bg \right\rangle + \psi_{gb} \left| gb \right\rangle + \psi_{gg} \left| gg \right\rangle$?

(g) If a measurement of the observable $\underline{\hat{F}}^1$, introduced in part (e), is made on the state

$$\left| \beta \right\rangle = \frac{1}{\sqrt{2}} \left(\left| bg \right\rangle + \left| gb \right\rangle \right)$$

find out the probability of getting the outcome $-\kappa$ and the final state. What would your findings be if the initial state was

$$\left| \beta \right\rangle = \beta_1 \left| bg \right\rangle + \beta_2 \left| gb \right\rangle$$

where $\left| \beta_1 \right|^2 + \left| \beta_2 \right|^2 = 1$?

2. A particle of mass m is confined in a 3-dimensional box having rectangular dimensions a, b and c. Find out the energy eigenvalues and eigenfunctions. Investigate the degeneracy of the eigenvalues if $a = b = c$.

(Hint: The vector space where the states of a 3-dimensional particle live is the tensor product space $V^x \otimes V^y \otimes V^z$ of the vector spaces associated with the three spatial dimensions x, y and z. Now for any operator

$$\hat{A} = \hat{A}_x + \hat{A}_y + \hat{A}_z$$

where $\hat{A}_x, \hat{A}_y$ and $\hat{A}_z$ are the extensions of operators acting on V^x, V^y and V^z respectively to $V^x \otimes V^y \otimes V^z$, if the eigenvalue equations

$$\hat{A}_x |A_x\rangle = A_x |A_x\rangle$$
$$\hat{A}_y |A_y\rangle = A_y |A_y\rangle$$
$$\hat{A}_z |A_z\rangle = A_z |A_z\rangle$$

are obeyed respectively in V^x, V^y and V^z (using the same notation to denote the original operators before extension), then

$$\hat{A} (|A_x\rangle |A_y\rangle |A_z\rangle) = (A_x + A_y + A_y)(|A_x\rangle |A_y\rangle |A_z\rangle)$$

holds in $V^x \otimes V^y \otimes V^z$.)

3. Consider a particle of mass m moving under a potential energy $V(x) = 1/2(K_x x^2 + K_y y^2 + K_z z^2)$ where K_x, K_y and K_z are positive constants.

   (a) Solve the eigenvalue problem for the energy of the system by using separation of variables, and showing that the problem reduces to three 1-dimensional harmonic oscillators.

   (b) In the light of the tensor product space formulation, the system can be reinterpreted as a composite system of three independent harmonic oscillators (oscillating with the same frequency). Reproduce the spectrum and the eigenfunctions of the Hamiltonian.

   (c) Obtain an expression for the degeneracy of the energy levels.

4. The hydrogen atom is surely an interacting composite system where a proton 'p' and an electron 'e' are bound by the coulomb interaction. Taking both the centre of mass and relative coordinate sectors into account, the full energy eigenstates would look like

$$\Psi_{n,l,m}\left(\vec{P}\right) = A\,\psi_{n,l,m}\left(r, \theta, \phi\right) \exp\left(\frac{i}{\hbar}\vec{P}.\vec{R}\right)$$

Here $\vec{r} = \vec{r}_e - \vec{r}_p$ is the relative coordinate, $\vec{R} = (m_e\vec{r}_e + m_p\vec{r}_p)/(m_e + m_p)$ is the centre of mass coordinate, $\vec{P}$ is the corresponding momentum and A is some constant. Is the state entangled? Argue that we can envisage the hydrogen atom formally as a non-interacting system of two fictitious particles of mass $M = m_e + m_p$ and $\mu = (m_e m_p)/(m_e + m_p)$. What would you say about the entanglement in this description?

5. Imagine a composite system comprising two *spin-half* particles[16]. Also assume that spin is the only attribute of these particles; they have no other observables[17].

---

[16] Refer to the spin-half problem (problem 7) following the section on angular momentum.
[17] This exercise is a simple example of something that is referred to as *addition of angular momenta* in QM.

(a) Write down the state space for this composite system using the spin-z basis for both particles.

(b) In the state space of the composite system, define the new observables

$$\hat{J}_i = \hat{J}_{1i} + \hat{J}_{2i}$$

where $i = x, y, z$. Here the suffixes, 1 and 2, refer respectively to the first and the second particles, and all the single particle operators are assumed to be extended to the state space of the composite system. Show that these operators obey the angular momentum commutation relations.

(c) Without using brute force, determine the possible eigenvalues of $\hat{J}_z$ and identify the corresponding eigenspaces.

(d) Instead of using the CSCO $\left\{\hat{J}_1^2, \hat{J}_2^2, \hat{J}_{1z}, \hat{J}_{2z}\right\}$ one can use the CSCO $\left\{\hat{J}_1^2, \hat{J}_2^2, \hat{J}^2, \hat{J}_z\right\}$[18]. Construct the transformation matrix connecting the corresponding bases: express the basis elements that are labelled by the eigenvalues of the operators of the second CSCO in terms of the basis whose members are labelled by the eigenvalues of the operators of the first.

6. The electron in the hydrogen atom is a *spin-half* particle.

(a) What do you imagine will be the state space of the hydrogen atom if we take electron spin into account. Write down a formal expression of a wave function in your proposed state space.

(b) Inspect how the degeneracy of the energy eigenvalues will be affected (recall that the previous degeneracy was n^2, with n being the principal quantum number).

Identical Particles

In this section we will finally have occasion to take up a class of quantum systems that have *no classical analogues*[19]. The setting of the discussion to follow, are composite systems made out of an assembly of elementary building blocks that we shall *call* "particles". Now in nature, as one would perhaps

---

[18]Note that here the operators $\hat{J}_1^2$ and $\hat{J}_2^2$ are actually redundant since both particles are spin-half.

[19]The content of this section should, rightfully, belong to the earlier section. But owing to its importance and generality we have decided to dedicate a full section to it.

imagine, there exists composite systems that are made out of elementary building blocks (i.e., particles) that are *identical*[20]. It turns out that this aspect of having exactly similar building blocks have rich and intriguing consequences that are actually observable. We shall lay down in this section a scheme for specifying the quantization rules that one must use for such systems.

Classification of identical particles

The first task in the quantization process is the specification of the state space. It turns out that within the category of systems that we are now considering, there are two broad subclasses of systems for which state spaces are intrinsically different. We refer to these subclasses in terms of the particles that constitute them. These particles are, therefore, called by different names. Let us describe this in more detail.

The specification of the state space involves two steps:

1. Specification of *one particle states*: One considers a quantum system consisting of just one unit of the elementary building block (i.e., a single-particle system). The members of the state space of a single particle are called **one particle states**.

2. Specification of the full state space: One assumes that there exists an **occupation number operator** (sometimes referred to, simply, as a *number operator*) corresponding to every one particle state. The spectra of these operators consists of nonnegative integers representing the number of particles *occupying* the respective one particle states. The CSCO for the state space is the set of all occupation number operators associated with all the one particle states corresponding to some basis of the one particle state space.

There are only two classes of identical particle systems that we see in nature. Correspondingly, there are only two species of identical particles that exist in our universe. The state space of the first kind have number operators whose spectra consists of the two integers, 0 and 1. Such particles are called **Fermions**[21]. The state space of the second kind have number operators whose spectra are the set of all nonnegative integers. These particles are called **Bosons.**

The spectra of the number operators are, of course, constrained by the total number of particles of the multiparticle system under consideration. However, there exists situations where the total number of particles are taken to be infinite.

---

[20]For a more exhaustive treatment of "identical particles", see (Cohen-Tannoudji et al. 1977).

[21]This is just the *Pauli's exclusion principle* in our new language.

Explicit description of the state spaces

Let us consider the one particle states to inhabit a state space spanned by a basis $\{|\psi_i\rangle \,; i = 1, 2, \ldots, g\}$ comprising g one particle states $|\psi_i\rangle$. We also leave open the possibility that the number g is infinite.

A basis of the multiparticle state space of a system comprising N identical particles can be taken to be the set of states $|n_1, n_2, \ldots n_g\rangle$ where n_j is an eigenvalue of the number operator $\hat{N}_j$ corresponding to the j-th one particle state $|\psi_j\rangle$. The eigenvalue n_j represents the number of particles occupying the state $|\psi_j\rangle$.

- *For a system of fermionic particles, the occupation number n_j can take values from the set $[0, 1]$.*

- *For a system of bosonic particles, the occupation number n_j can take values from the set $[0, 1, 2, \ldots, N]$.*

Symmetric and antisymmetric spaces

Although we have, in our description, used the occupation number representation to specify the state space of identical particles, we would like to mention that a different strategy was also possible[22].

In this approach, one starts from a one particle state space U, as usual, having a basis $B^U = \{|\psi_i\rangle \,; i = 1, 2, \ldots, g\}$ (assumed to be g-dimensional, for concreteness). To describe a system of N particles, one then constructs, as an intermediate step, a tensor product space V of the state spaces of the N particles, pretending that they are *distinguishable*. The state spaces of all the N particles are assumed to be identical. A basis B^V of the multiparticle state space V comprises the set of tensor product states $\left|\psi_\alpha^{(1)}, \psi_\beta^{(2)}, \ldots, \psi_\kappa^{(N)}\right\rangle$ of N one particle states $\left|\psi_\gamma^{(j)}\right\rangle$ where the index j denotes the particle, and the index γ indicates a one particle state belonging to B^U. Every state $|\psi\rangle$ in V can then be written as

$$|\psi\rangle = \sum_K C_K \left|\psi_\alpha^{(1)} \psi_\beta^{(2)}, \ldots, \psi_\kappa^{(N)}\right\rangle$$

where K stands for a particular choice of one particle states belonging to B^U, in which the particles are put. Thus K is essentially an N-tuple comprising the values of the indices $\{\alpha, \beta, \ldots, \kappa\}$; each of the indices, in turn, can run over the set of indices $(1, 2, \ldots, g)$ labelling the states in B^U.

Now a permutation of the particles in $|\psi\rangle$ amounts to permuting the sequence of the N integers: $(1, 2, \ldots, N)$ in the superscripts of all the basis states of B^V in the expansion of $|\psi\rangle$.

---

[22]This has actually been the historical route, and it is still very relevant and important.

The set of states $|\psi\rangle$ which remain invariant under an arbitrary permutation of the particles forms a subspace of V. Such states are called **symmetric states**, and the subspace is called the **symmetric subspace**. We shall denote this subspace by S. An example of a symmetric state, for $N = 2$, is[23]

$$\left|\psi_i^{(1)}\psi_j^{(2)}\right\rangle + \left|\psi_j^{(1)}\psi_i^{(2)}\right\rangle.$$

The set of states $|\psi\rangle$ which become $\varepsilon|\psi\rangle$ under a permutation, where $\varepsilon = +1$ or -1 depending on whether the permutation is even or odd, also forms a subspace. Such states are called **antisymmetric states**, and the subspace is called the **antisymmetric subspace**. We shall denote this subspace by A. An example of an antisymmetric state, for $N = 2$, is[24] $\left|\psi_i^{(1)}\psi_j^{(2)}\right\rangle - \left|\psi_j^{(1)}\psi_i^{(2)}\right\rangle.$

It turns out that the symmetric subspace is precisely the state space of Bosons, and the antisymmetric subspace is precisely the state space of Fermions.

One proceeds with the quantization by imposing the constraint that

physical states of identical particles must belong to either the symmetric subspace or the antisymmetric subspace.

Sometimes, this is referred to as the **symmetrization postulate**.

A *natural* basis for the symmetric and the antisymmetric subspaces can be constructed as follows:
First of all we observe that every basis vector in B^V corresponds to a selection $L = \{|\psi_1\rangle, |\psi_2\rangle, ,\ldots, |\psi_l\rangle\}$ of l basis vectors in B^U that have been populated (clearly, $1 \leq l \leq N$). It is possible to assign a set of l occupation numbers $\{n_1, n_2, \ldots, n_l\}$ with n_i denoting the number of particles in state $|\psi_i\rangle \in L$. More systematically, one can define an ordered set $(n_1, n_2, \ldots, n_g)$, which, for brevity, we denote by the vector $\vec{n}$, such that we have an occupation number entry for every one particle state in the basis B^U. For the one particle states that are not in the chosen set L, we simply assign an occupation number zero. It is clear that a specification of $\vec{n}$ does not uniquely determine a basis element of B^V. There will, in general, be several vectors in B^V corresponding to a given $\vec{n}$ who are related to each other by permutations.
One can construct a vector $\left|\vec{n}^S\right\rangle$ by

$$\left|\vec{n}^S\right\rangle = \frac{1}{N!}\sum_I P_I \left|\psi_\alpha^{(1)}\psi_\beta^{(2)}, \ldots, \psi_\kappa^{(N)}\right\rangle$$

---

[23]Note that, here the states $\left|\psi_k^{(i)}\right\rangle$ are arbitrary states in V. They need not correspond to the basis states in B^U. Here we assume $i \neq j$; the case $i = j$ is trivially symmetric.

[24]Here again, the states $\left|\psi_k^{(i)}\right\rangle$ are arbitrary states in V. Note that For $N = 2$, there are no nontrivial even permutations. Here also, we assume $i \neq j$. Can you see what will happen if $i = j$?

where $I = (i_1, i_2, \ldots, i_N)$ is a permutation of the set $(1, 2, \ldots, N)$ and P_I is a permutation operator that acts on a basis vector of B^V according to

$$P_I \left| \psi_\alpha^{(1)} \psi_\beta^{(2)}, \ldots, \psi_\kappa^{(N)} \right\rangle = \left| \psi_\alpha^{(i_1)} \psi_\beta^{(i_2)}, \ldots, \psi_\kappa^{(i_N)} \right\rangle$$

Note that, in the above formulae, the label $\psi_\gamma^{(j)}$ designates the single particle state $|\psi_\gamma\rangle$ to j-th particle but such single particle states need not be distinct for different j. It is clear that for every I, the vector $P_I \left| \psi_\alpha^{(1)} \psi_\beta^{(2)}, \ldots, \psi_\kappa^{(N)} \right\rangle$ corresponds to the same $\overrightarrow{n}$ as $\left| \psi_\alpha^{(1)} \psi_\beta^{(2)}, \ldots, \psi_\kappa^{(N)} \right\rangle$. Moreover every basis vector that corresponds to this $\overrightarrow{n}$ is generated by some P_I in this way. It is easy to see that the vector $\left| \overrightarrow{n}^S \right\rangle$ is symmetric[25]. The above properties justify the notation $\left| \overrightarrow{n}^S \right\rangle$ (the superscript S stands for symmetric and $\overrightarrow{n}$ specifies the occupation distribution). The set of vectors $B^S = \{ \left| \overrightarrow{n}^S \right\rangle | \sum_{i=1}^g n_i = N \}$ forms a basis of the symmetric subspace S.

For the antisymmetric subspace, the appropriate superposition turns out to be

$$\left| \overrightarrow{n}^A \right\rangle = \frac{1}{N!} \sum_I \varepsilon_I P_I \left| \psi_\alpha^{(1)} \psi_\beta^{(2)}, \ldots, \psi_\kappa^{(N)} \right\rangle$$

where $\varepsilon_I = +1$ or -1 for even and odd permutations respectively. The introduction of ε_I ensures that $\left| \overrightarrow{n}^A \right\rangle$ is antisymmetric. It is to be noted that in this case, unless the number l of one particle states that we choose to populate, is exactly N, the vector $\left| \overrightarrow{n}^A \right\rangle$ will vanish (i.e., it will become the additive identity vector, as you shall prove in the problem set shortly). This also means that the occupation numbers n_i are restricted to 0 and 1. The set of vectors $B^A = \{ \left| \overrightarrow{n}^A \right\rangle | \sum_{i=1}^g n_i = N \}$ forms a basis of the antisymmetric subspace A.

It turns out that the basis vectors $\left| \overrightarrow{n}^S \right\rangle$ and $\left| \overrightarrow{n}^A \right\rangle$ in B^S and B^A can be identified with the basis vectors $|n_1, n_2, \ldots, n_g\rangle$ for Bosons and Fermions defined earlier in terms of occupation numbers.

Observables of the multiparticle system

The final step in the quantization procedure, namely the specification of representations of all observables of interest will depend on the specifics of the actual system under consideration, and therefore, cannot be written down in general. However, the curious reader will surely wonder how one can even guess the form of the observables of interest for identical particle systems. Let us provide a rough idea of how this can be done.

One can imagine in the full tensor product space (that would be used as the state space of a system of distinguishable particles), operators which can be constructed by taking clues from a classical system along the lines that was discussed earlier. These operators would include the extensions of the one

---

[25]The factor $1/N!$ is for computational convenience, as you will discover in the problem set.

particle operators (for different particles) to the full tensor product space. Now, there exists a subclass of such operators, called **symmetric operators**, which remain invariant in form under arbitrary permutation of the particle index associated with the individual component operators (which are the extensions of one particle operators associated with the different particles). It turns out that the symmetric and the antisymmetric subspaces are invariant under the action of symmetric operators so that the symmetric operators are well defined in the symmetric and antisymmetric subspaces. In the occupation number representation, these operators can be transcribed in terms of a new set of operators which are appropriate to this description. Generally, one defines ladder operators on the occupation number states $|n_1, n_2, \ldots n_r\rangle$, and then attempts to write the symmetric observables discussed above in terms of the ladder operators, somewhat along the same lines that was followed for the harmonic oscillator.

Exchange degeneracy

In the opening paragraph of this section we said that indistinguishability has observable consequences, which is why we had to deviate from our usual tensor product prescription for the quantization of systems of identical particles. It is finally time to demonstrate this, and justify our choice of quantization.

Imagine a system of two identical particles whose state space V is constructed from the one particle state space U. We shall attempt a quantization of this system using the tensor product prescription (without symmetrization or number operators) and see what it implies.

Let us assume that U corresponds to a 2-level system and that $B = \{|\chi_1\rangle, |\chi_2\rangle\}$ is a basis in U. Consider that the system is in an initial state such that one of the particles is in a state $|\chi_1\rangle$ and the other particle is in the state $|\chi_2\rangle$. The most general state $|\psi\rangle$ of this kind will be the superposition

$$|\psi\rangle = c_1 \left|\chi_1^{(1)}\right\rangle \left|\chi_2^{(2)}\right\rangle + c_2 \left|\chi_2^{(1)}\right\rangle \left|\chi_1^{(2)}\right\rangle$$

where c_i ($i = 1, 2$) are complex coefficients, and the superscripts denote the particles. Now consider a general state $|\xi\rangle$ in U. Obviously

$$|\xi\rangle = d_1 |\chi_1\rangle + d_2 |\chi_2\rangle$$

where d_1 and d_2 are again complex coefficients. Now consider, in V, a state

$$|\phi\rangle = \left|\xi^{(1)}\right\rangle \left|\xi^{(2)}\right\rangle$$

The probability $P[|\psi\rangle \to |\phi\rangle]$ of the event $|\psi\rangle \to |\phi\rangle$ will then be given by

$$P[|\psi\rangle \to |\phi\rangle] = |\langle\phi|\psi\rangle|^2$$

It is easy to check that

$$P\left[|\psi\rangle \to |\phi\rangle\right] = |d_1|^2 |d_2|^2 |c_1 + c_2|^2$$

Evidently, the probability depends on the coefficients c_i. This means that our description makes a clear distinction between particles 1 and[26] 2. The probability depends on the weight of *which particle is in which one particle state* in the superposition that constitutes the initial state $|\psi\rangle$. This is never observed in reality. Every preparation of the state $|\psi\rangle$ (with one of the particles in state $|\chi_1\rangle$ and the other particle in state $|\chi_2\rangle$) will always give the same probability. In other words, it is impossible to prepare two or more *distinct* physical states compatible with the condition that one of the particles is in a state $|\chi_1\rangle$ and the other particle is in the state $|\chi_2\rangle$. In fact, it is precisely for this reason that the particles are called identical in the first place.

The feature of having different possible physical states (leading to different observable outcomes) under exchange of identical particles is known as *exchange degeneracy*[27].

In the correct description, exchange degeneracy must be absent.

It is not difficult to see that if the initial and final states are appropriately symmetrized, then the exchange degeneracy is removed.

Problems

1. The state space of a system of N identical particles is constructed from a one particle state space U having dimension g. Determine the dimension of the symmetric and antisymmetric state spaces S and A.

2. Show that the permutation operators form a group[28].

3. If we have a finite set $G = \{g_1, g_2, \ldots, g_n\}$ that forms a group under a group operation '$\star$', then according to the, so called, *rearrangement theorem* in group theory, $g_k \star G = G$ where[29] $g_k \star G = \{g_k \star g_i \,;\, i = 1, 2, \ldots, n\}$.

---

[26]Note that the dependence on d_1 and d_2 is irrelevant in this context. It merely specifies the choice of the state $|\xi\rangle$ in the one particle state space. In this example, the final state $|\phi\rangle$ is invariant under an exchange of the particles 1 and 2.

[27]In the above discussion the exchange degeneracy was due to the initial state. In the general case, it can, of course, be present in the final state as well. In fact, this artifact of the present example (i.e., absence of exchange degeneracy in the final state) somewhat masks the effect of exchange degeneracy: under an exchange of particles, although we have a distinct initial state, the final probability remains unaltered.

[28]The definition of a *"group"* was introduced in chapter 3 (footnote 20).

[29]This is easy to prove and you should actually prove it.

(a) If $\hat{P}_I$ are the permutation operators on V (in the notation used in the text), using the above theorem, show that the operators

$$\hat{S} = \frac{1}{N!} \sum_I \hat{P}_I \quad \text{and} \quad \hat{A} = \frac{1}{N!} \sum_I \varepsilon_I \hat{P}_I$$

are projection operators[30].

(b) Show that $\hat{S}$ and $\hat{A}$ are actually projections onto the symmetric and antisymmetric subspaces. For this reason they are called the *symmetrizer* and *antisymmetrizer* respectively.

(c) Prove that the symmetric and antisymmetric subspaces S and A are orthogonal.

(You will probably need the following fact: A permutation operator that exchanges exactly two particles keeping the others unaffected is called a *transposition operator* or, sometimes, *exchange operator*. Every permutation operator can be expressed as a product of transposition operators. Although the *factorization is not unique*, the number of such transposition operator factors is unique for a given permutation. It is called the *parity* of the permutation. In a permutation group, exactly half the elements have *even* parity and half have *odd* parity[31].)

(d) For $N = 2$, write down the form of the projectors $\hat{S}$ and $\hat{A}$. Show that they are also *supplementary*: $\hat{S} + \hat{A} = \hat{I}$ (identity operator)[32]. Is this true for $N > 2$?

4. Consider a system of 3 particles, and a set of one particle states $B = \{|\psi_1\rangle, |\psi_2\rangle, |\psi_3\rangle, |\psi_4\rangle\}$.

(a) Write down a Bosonic state if we choose to populate exactly (i) three (ii) two (iii) one state(s) from the set B.

(b) Show that, if you choose to populate three states, a Fermionic state is proportional to $\det F$, where F is a matrix defined by[33] $F_{ij} = |\psi_j^{(i)}\rangle$ (where i denotes the particle, and j the one particle state assigned to it). What do you expect to see if you choose to populate less than three states[34]?

---

[30] Check out the section of projection operators in appendix 'B'. The proofs should make it clear why the factors $1/N!$ have been hanging around.

[31] This is a general permutation group result and you are invited to prove it as well.

[32] Another way of saying this is that the state space is a *direct sum* of the symmetric and antisymmetric subspaces. Check out appendix 'B' for the meaning of "direct sum" if you haven't done it as yet.

[33] Incidentally, $\det F$ is called the *Slater determinant,* and this formula actually holds for any N.

[34] You have again stumbled upon Pauli's exclusion principle!

(c) Assuming the set B to be orthonormal, normalize the states in part (a) and (b).

(d) Generalize the argument in part (c) to obtain general expressions for the normalization constants for the Bosonic and Fermionic basis vectors $\left|\vec{n}^{S}\right\rangle$ and $\left|\vec{n}^{A}\right\rangle$ that were identified with basis vectors in occupation number representation.

5. Construct an argument to convince yourself that symmetrization removes exchange degeneracy.

Appendix A

Probability

Since QM is a probabilistic theory, it is not surprising that the term "probability" has occurred all over the place in this book. To have a concrete understanding of the content of this book, a reasonably precise understanding of *probability* is essential. Now the word probability comes with various connotations in everyday usage, and there have been attempts to mathematically formalize and interpret probability in various ways to make it as inclusive of those connotations as possible. In this appendix, we shall give a brief discussion on the interpretation of probability that is relevant to the standard view of QM. This is known as the *frequentist* interpretation and is by far the most common interpretation used in physics.

Preliminary Background

In this section, we introduce the basic concepts and definitions which are essential to any discourse on probability.

Experiment *An experiment is any process that has associated with it a well defined set of outcomes.*

In the frequentist interpretation, we shall always assume that an experiment is infinitely repeatable.

Simple examples are: 1) tossing a coin, where the outcomes are head or tail, and 2) rolling a die, where the outcomes are the pips. It is to be noted that what constitutes an outcome depends on the context and the problem under study. The outcomes need only be distinguished and characterized by what is of interest to the experimenter. For example, we can imagine a shooting experiment designed to study the aim of the shooter. We would consider the outcomes to be the distances from the bulls eye where the bullet hits the target board. There might be other aspects of the output that also vary from one experiment to another (e.g., the depth of the dent produced by the bullet) which are not relevant.

Sample Space *The set of all possible outcomes of an experiment is called the sample space of the experiment.*

The sample space for the "coin tossing" experiment would be the set: {*head, tail*}. Similarly, for the "die rolling" experiment the sample space would be the set of pips: $\{1, 2, 3, 4, 5, 6\}$. If we imagine an experiment where two dice are rolled simultaneously, we could represent the sample space by the set of ordered pairs $\{(1, 1), (1, 2), \ldots, (6, 6)\}$ where the first entry of an ordered pair is the outcome from the first die and the second entry is that from the second.

Event Space[1] *An **event** is characterized by a subset of the sample space. The set of all possible events is referred to as the event space.*

For a discrete sample space, the set of all subsets, called the *power set,* is thus an event space containing all possible events.

If A and B are two events then the event characterized by the intersection set $A \cap B$ is said to be a **conjunction** of the two events and is referred to as *A **and** B*. The event characterized by the union $A \cup B$ is called a **disjunction** of A and B and the event is referred to as *A **or** B* .

Two events A and B are **mutually exclusive** when the corresponding subsets have a null intersection: $A \cap B = \phi$.

For an event A, its **negation** is characterized by the complement $\bar{A}$, and it is referred to as **not** *A*.

If a subset of the sample space is a singleton (i.e., it has just one element), it characterizes a **simple** or an **elementary** event. A non-empty subset that is not a singleton characterizes a **compound** or a **non-elementary** event. Thus, every possible outcome of an experiment corresponds to a *simple* event. On the other hand, the occurrence of one or the other of a well defined subset of outcomes corresponds to a *compound* event that is characterized by the subset. So a compound event is one which can happen in several different *ways,* where each *way* corresponds to an elementary event.

Let us illustrate these definitions with examples. In the experiment where two dice are rolled, a particular outcome, say $(2, 5)$, corresponds to a simple event. If, for some reason, we are interested in the situation where the outcomes of the two die rolls add up to four, then we are talking about a compound event which would be characterized by the subset $\{(1, 3), (2, 2), (3, 1)\}$ of the sample space. In the single die rolling experiment, a pip $\{3\}$ is a simple event while if we talk about the event that the "pips must be even", it would

---

[1]What we discuss here is not the most general scenario. Please see the last section for a more complete definition.

be a compound event described by the subset $\{2, 4, 6\}$. The event that the pip must be "even *and* less than 5" would be the set $\{2, 4\}$, which is the intersection $\{2, 4, 6\} \cap \{1, 2, 3, 4\}$. Again, if we consider the event that pip must be "even *or* less than 5", it would be described by the set $\{1, 2, 3, 4, 6\}$, which is given by the union $\{2, 4, 6\} \cup \{1, 2, 3, 4\}$. The set of events $\{$"pip is even", "pip is odd"$\}$ are mutually exclusive: $\{2, 4, 6\} \cap \{1, 3, 5\} = \phi$.

Frequentist Interpretation

Now we are in a position to introduce the definition of *probability*. A probability will always be associated with an *event* which in turn is tied to an *experiment*.

Probability of an event *The ratio of the number of times that an event occurs to the total number of experiments, in the limit when the experiment is repeated an infinite number of times, is defined to be the probability of the event.*

If n_A denotes the number of times that an event A occurs, and N represents the total number of experiments, then the probability $P[A]$ is defined by

$$P[A] = \lim_{N \to \infty} \frac{n_A}{N}$$

Thus, the probability of an event in the frequentist view is given by the *frequency* of occurrence of the event in the *long run*.

The probability $P[x]$ of the simple events $\{x\}$ where x is an element of the sample space, defines what is known as the ***probability mass function***[2]. Note that this function is defined on the sample space and not on the event space. The probability of an arbitrary event $P[A]$ is given in terms of the probability mass function as

$$P[A] = \sum_{x \in A} P[x]$$

---

[2]Remember, that this is in the discrete case. In the case of an uncountable sample space, it is not even necessary that the probability of an elementary event be defined. It is only required that probabilities can be assigned to subsets of the sample space that constitute the event space.

Properties

We now list a set of properties that follow easily from the above definition of probability.

1. The probability of any event A lies between zero and one:

$$0 \le P[A] \le 1$$

2. The **total probability** $P[S]$, which is the probability that some outcome in the sample space S will occur, is unity.

$$P[S] = \sum_{x \in S} P[x] = 1$$

3. The probability of a compound event comprising a countable collection (possibly infinite) of mutually exclusive events, characterized by disjoint subsets A_i (where $i = 1, 2, \ldots n$ with the possibility $n \to \infty$ being allowed) of the event space, is equal to the sum of the probabilities of the disjoint events[3] :

$$P[\cup_{i=1}^{n} A_i] = \sum_{i=1}^{n} P[A_i]$$

4. The probability $P[\bar{A}]$ of the event $\bar{A}$, that some event characterized by A *does not* occur, is given by

$$P[\bar{A}] = 1 - P[A]$$

Here $\bar{A}$ is the complement of A.

5. The probability $P[A \cup B]$ of a compound event $A \cup B$, that an event A *or* an event B occurs, is given by

$$P[A \cup B] = P[A] + P[B] - P[A \cap B]$$

Random variables

A **random variable** provides a representation of the sample space in terms of real numbers. Formally, the random variable is a function from the sample space to the set of real numbers[4]:

$$X : S \longrightarrow \mathbb{R}$$

The range of the random variable is often called the **state space**.

---

[3]More precisely, the requirement is *pairwise* disjointness. The subtle difference between pairwise disjointness and ordinary disjointness can be ignored for most practical purposes.

[4]More generally, it is a function to a *measurable* space. For our purpose, it will be adequate to restrict the definition to the special case of real numbers.

It is often the case that one is interested in the statistics and probability of various quantities, which are all dependent on (i.e., functions of) a few basic quantities represented by a few basic variables, often called coordinates. In this situation it is clearly uneconomical and unnecessary to specify the probabilities of all the different quantities. It is adequate to specify the probability of the basic variables. It then makes sense to call the set from which the basic variables takes their values, the *state space*[5]. In most cases of interest, the basic variables are real valued.

In general, it might be necessary to use more than one real number to represent a point in the sample space. In this case, we call the random variable **multivariate** (as opposed to **univariate**), and the random variable is then defined by a function

$$X : S \longrightarrow \mathbb{R}^n$$

The value of this function, the multivariate random variable is a real n-tuple, and it is often called a **random vector**.

To give an example of random variables, we can imagine an experiment involving darts thrown at a dart board. The sample space would naturally be the set of points on the dart board. A natural choice of the random vector in this case could be the coordinates (x, y) with respect to a 2-dimensional Cartesian coordinate system set up on the dart board.

For a discrete sample space, the random variable can be trivially chosen to be any appropriate set of integers. It is seldom necessary to explicitly invoke the random variable for discrete sample spaces.

Once a random variable has been chosen to represent the sample space, for all practical purposes, one can forget about the underlying sample space.

For the rest of this section we shall continue to denote the sample space by S and the associated random variable by X.

Probability density function

It is obvious, that when the sample space is uncountable, it is impossible to assign a probability function to the event space through the probability mass function (since a *sum* can no longer be performed on an uncountable set). As one would imagine, one needs an integral in such cases. To this end one often uses a *probability density function*.

A **probability density function** (**PDF**) $p(x)$ is a continuous, nonnegative function defined on the codomain $\mathbb{R}$ of the random variable, satisfying the condition

$$\int_{-\infty}^{+\infty} p(x)\, dx = 1$$

---

[5]For example, in classical mechanics, the dynamical variables are all functions of the positions and momenta of the particles that make up the system, collectively called the *state* of the system.

The PDF allows one to assign probabilities to **intervals** (subsets) $\mathbb{A} = [a, b]$ of $\mathbb{R}$:

$$P(a \leq x \leq b) = \int_a^b p(x)\, dx$$

Now the events A can be defined as preimages of intervals of $\mathbb{R}$ with respect to the random variable function X:

$$A = \{s \in S \; : \; X(s) \in \mathbb{A}\}$$

Hence, the probability $P[A]$ associated with the event A is defined by

$$P[A] = P(\mathbb{A})$$

Thus $P(\mathbb{A})$ can be used as a shorthand for expressing the probability of the event A:

$$P(\mathbb{A}) \equiv P[\{s \in S : X(s) \in \mathbb{A}\}]$$

Cumulative distribution function

One often works with what is known as a *cumulative distribution function* instead of the density function.

A *cumulative distribution function* (**CDF**) $P_X(x)$ is a function defined on the range $\mathbb{R}$ of a random variable X such that the value of the function at a point $x \in \mathbb{R}$ is the probability of the event that the value of the random variable is not greater than x:

$$P_X(x) = P(X \leq x)$$

One can associate probabilities with half open intervals $(a, b]$ in the range of the random variable using cumulative functions

$$P(a < x \leq b) = P_X(b) - P_X(a)$$

A CDF can be associated with a PDF as

$$P_X(x) = \int_{-\infty}^x p(t)\, dt$$

and conversely, the PDF is given in terms of the CDF by the derivative:

$$p(x) = \frac{dP_X(x)}{dx}$$

Since for a continuous random variable, the probability at any particular point is zero, it follows that

$$P(a < x \leq b) = P(a \leq x \leq b) \equiv P[a, b]$$

Any CDF is a *right-continuous, increasing* function defined on $\mathbb{R}$ satisfying the conditions[6]

1. $\lim\limits_{x \to \infty} P_X(x) \longrightarrow 1$

2. $\lim\limits_{x \to -\infty} P_X(x) \longrightarrow 0$

Unlike a PDF, a CDF may be easily defined also for a discrete random variable using the probability mass function (in this case defined on the range of the random variable). If the random variable takes values from a set $\{x_1, x_2, \ldots\}$, then the CDF will be given by

$$P_X(x) = \sum_{x_i \leq x} P(X = x_i) = \sum_{x_i \leq x} p(x_i)$$

Here $p(x_i)$ is the probability mass function.

In the literature the usage of the term **probability distribution** is a bit ambiguous. Most of the time, it is used to refer to the probability density functions, but, often, it is also used to refer to cumulative distribution functions. Sometimes, it is also used as a blanket term that includes probability mass functions. It has to be understood from the context how the term is being used.

Conditional probability

The **conditional probability** $P(A \mid B)$ is the probability for an event A to occur given that another event B has occurred. It is *defined* by

$$P(A \mid B) = \frac{P(A \cap B)}{P(B)}$$

It is worth remarking that although the conditional probability is defined in terms of probabilities of certain events in the event space, there is *no event* (subset of sample space) in the event space that could be called a *conditional event* to which the conditional probability can be associated. If one insists, the conditional event '$A \mid B$' can be interpreted as an event in a *redefined* (*restricted*) sample space which is the event B (see Figure A.1).

To see an example of conditional probability, let us go back to our die rolling experiment and assume that all the pips have equal likelihood. The conditional probability that a die roll yields $\{2\}$ given that the outcome is even, is $(1/3)$. With the possibilities restricted to even outcomes $\{2, 4, 6\}$, the event $\{2\}$ is just one of the three equally likely possibilities. Using the above formula

$$P(2 \mid \text{even}) = \frac{P(2 \cap \text{even})}{P(\text{even})} = \frac{1/6}{1/2} = \frac{1}{3}$$

---

[6]It turns out that these conditions are also sufficient for a function to qualify as a CDF.

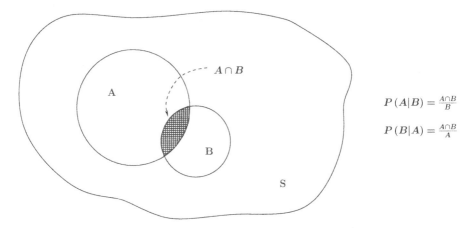

$$P(A|B) = \frac{A \cap B}{B}$$

$$P(B|A) = \frac{A \cap B}{A}$$

FIGURE A.1: Conditional Probability: Observe that there is no subset of the sample space S that can be assigned to a *conditional event*. However, the intersection $A \cap B$ can be taken to be the conditional event $(A|B)$ (or $(B|A)$) if we imagine a restricted sample space B (or A). Also note that, in general, $P(A|B) \neq P(B|A)$.

It is easy to imagine innumerable real life scenarios where conditional probabilities are relevant. For example, in an outbreak of an epidemic, a clinician may be interested in the probability that a person who is known to have contracted the disease will actually test positive under a certain detection test. This would give the reliability of the test indicating the propensity of the test to produce false negatives. On the other hand a patient, who has tested positive, could be interested to know how likely it is that she has actually contracted the disease given that she has got a positive test result. This will depend on the propensity of the test to produce false positives.

It should be noted that conditional probability is clearly not symmetric: $P(B \mid A) \neq P(A \mid B)$. In the above example, it could be that the test under question detects a biomarker that is *only sometimes* elevated due to the disease but it is *almost never elevated in absence* of the disease. Such a marker will yield high false negatives but very few false positives. Consequently, the probability for a person having the disease to test positive may not be very high, but the probability for a person who has tested positive to actually have the disease will be high.

It is easy to see, from the definition of conditional probability, that it can be inverted. We have

$$P(B \mid A) P(A) = P(A \cap B) = P(A \mid B) P(B)$$

which implies

$$P(A \mid B) = \frac{P(B \mid A) P(A)}{P(B)}$$

This is known as **Bayes' formula** and it constitutes the basis of Bayesian statistics. In Bayesian statistics $P(A)$ is called the *prior* probability, $P(B \mid A)$ is called the *likelihood* and $P(A \mid B)$ is called the *posterior*. The denominator on the right hand side $P(B)$ is often called the *marginal* probability.

Independent events

The concept of conditional probability naturally leads to the notion of *independent events*. Two events A and B are said to be **independent events** when the conditional probability for A given B is the same as the absolute probability of A:

$$P(A) = P(A \mid B)$$

That is, the probability of A is unaffected by whether or not B occurs. Mathematically this is equivalent to the condition:

$$P(A \cap B) = P(A) P(B)$$

Evidently, the definition is symmetric so that, if A is independent of B, then B is also independent of A.

One should note that up until the introduction of conditional events, all events that we talked about (e.g., A and B, A or B, not A, mutually exclusive events, etc.) could be defined in terms of the elements of the event space. However, we had to invoke the concept of *probability* to introduce the definition of conditional events and the related idea of independent events.

Shortcomings of the frequentist view

The frequentist view of probability does not include many scenarios within its scope where the notion of probability or chance is still very much applicable. For example, if we wish to know the likelihood that "it will rain tomorrow", the question cannot even be posed within the frequentist interpretation of probability. Since there is just *one* tomorrow, the experiment is not repeatable. Indeed, there are myriads of instances where we talk of chance but the experiment in question cannot even be repeated *once,* let alone the possibility of repeating infinite times.

Since it is impossible, even in theory, to repeat an experiment an infinite number of times, and the frequentist interpretation does not give us any clue as to how high the number of experiments must be before the probability can be objectively inferred, the issue of convergence remains unsettled. There is always the possibility that at some large N (where N is the number of experiments), the frequency appears to converge, then it drifts away before finally damping to its supposed final value. One might imagine that it could be possible to incorporate, in the definition of probability, the fact that for finite N, the probability will be given up to some *error* with respect to the *true* probability. However, this will have the problem that any definition of error must invariably use some notion of probability leading to a *circularity* in the definition of probability!

Alternative Interpretations

There are two broad categories of interpretations called *physical* and *evidential* probability. The frequentist interpretation, that we have discussed, belongs to the "physical" probability category. Some of the other commonly used interpretations of probability are the *Classical* and *Bayesian* views. The classical view, again, belongs to the class of physical probability while the Bayesian view is the main candidate representing evidential probability interpretation.

In the classical view one conceives a class of equally likely base events (somewhat like the elementary events we discussed earlier) such that every other event can be associated with a number of favourable base events (which lead to it). The probability of an event is then defined to be the ratio of the number of favourable base events to the total number of possible base events. This, rather simplistic, view of probability obviously suffers from the disadvantage that it is limited only to those scenarios which admit an unbiased set of base events, which is quite often not the case in real life.

In the Bayesian approach, one adopts the view that probability is a *degree of belief* that a certain event will occur (or that a certain fact is true, etc.). These degrees of beliefs are constrained by consistency requirements (laws of probability[7]). The Bayesian method lays down rules on how the beliefs should be *updated* based on subsequently acquired data. Briefly, the method comprises the following:
A *prior* probability $P(H)$ for the hypothesis H under question is first generated based on available information. Then, in the light of subsequently acquired data D, the prior probability is *updated* by making use of Bayes' formula:

$$P(H \mid D) = \frac{P(D \mid H) P(H)}{P(D)}$$

to generate what is called the *posterior* probability $P(H \mid D)$. The step is then repeated by using the posterior as a prior. In this way, by repeated use of Bayes' formula, the posterior is updated to yield increasingly improved degrees of belief. The updation scheme has to make use of some process (mathematical model) that provides a *likelihood* for the acquired evidence (i.e., the data) to have been generated from the hypothesis under question. This likelihood is the conditional probability $P(D \mid H)$.

The main criticism held by the opponents of this view is that it is *inherently subjective*[8].

---

[7]These have been discussed in the next section.

[8]There is, however, a variant of the Bayesian interpretation called the objective Bayesian approach. As the name suggests, this view asserts that even Bayesian probabilities can be made objective. Essentially it extends the underlying logic system to accomplish this. In ordinary propositional logic, a proposition can either be true or false. In the generalized

Formal Description[9]

Although there is considerable amount of disagreement among experts as to what should be the correct interpretation of probability, people more or less agree on how a general probability should behave irrespective of the interpretation. These properties, which are shared by all the interpretations of probability, have been abstracted and generalized to develop a formal definition of probability.

The formal description of probability is not tied to any particular interpretation. It is a mathematical structure that has been motivated by the *important* properties of probability[10].

The formal description of probability defines a *probability space* comprising a *sample space*, an *event space* and a *probability function*[11].

Sample space

Since the formal description of probability is no longer attached to an interpretation, the sample space is no longer defined as the set of outcomes of some experiment (although this remains the underlying intended interpretation). Formally,

the sample space is defined as a non-empty set.

Event space

In the first section, we had defined the event space to be the power set of the sample space. Actually, it is neither essential nor possible, in general, to do this.

When the sample space forms a continuum (and is therefore uncountable), it is in general impossible to use the power set as an event space. To give an example, let us return to the "shooting" experiment. We could consider the sample space to be the set of real numbers that stand for the distance of the point from the bulls eye where the bullet hits the target board. If 60 cm be the radius of the target board, then the sample space S could be imagined

---

system, a proposition is allowed to take any value in the range $[0, 1]$. The aim of objectivists is to lay out a prescription for predicting the degrees of belief which would be so constrained that scope of subjectivity will be completely eliminated. Even a robot will come up with the same expectation.

[9]This section may be omitted on first reading. However, it provides a concrete example of a formal theory (without attached interpretations) that was introduced in the first chapter.

[10]The first major formalization was provided by Russian mathematician, Andrey Kolmogorov in 1933.

[11]More technically, called a *probability measure*. Formally, a probability is a *measure* and a *probability space* is a *measurable space* of measure one.

to be the real interval $[0, 60]$. An interval $[10, 15]$ would comprise a subset that would qualify as an event in the event space. However, if we consider the subset of all irrationals in the interval $[10, 15]$, that would not qualify as an event. It turns out that for such a set, one cannot associate a notion of *size* (in the mathematician's jargon, such a set is not *measurable*). This has the consequence that it is impossible to assign a probability (to be defined in the next section) to such subsets. In fact, it is also impossible, in general, to assign nonzero probabilities to arbitrary singletons of such a sample space.

In different contexts and scenarios, we may not be interested in all events associated with all possible subsets of the sample space. For example, in the die rolling experiment, if for some reason, the events of interest are the ones for which the pips belong to disjoint pairs of consecutive integers, then a perfectly acceptable event space would be

$$\{\{1, 2\}, \{3, 4\}, \{5, 6\}, \{1, 2, 3, 4\}, \{1, 2, 5, 6\}, \{3, 4, 5, 6\}, S, \phi\}$$

Here S is the full sample space $\{1, 2, 3, 4, 5, 6\}$ and ϕ denotes the empty set. We observe that for any set of events in this space, their *conjunction* and *disjunction* is also a member of the space. We also note that, for every event in the space, its *negation* is also a member of the space. Thus, in order to include events like $\{\{1, 2\} \ or \ \{3, 4\}\}$, the four-outcome sets such as $\{1, 2, 3, 4\}$ have been included. For any set of events, say $\{\{1, 2, 3, 4\}, \{3, 4, 5, 6\}\}$, the event $\{\{1, 2, 3, 4\} \ and \ \{3, 4, 5, 6\}\}$, which is $\{3, 4\}$, is also a member of the event space. The presence of the empty set ϕ allows for the description of the conjunction of mutually exclusive events like $\{\{1, 2\} \ and \ \{3, 4\}\}$. For every event, say $\{1, 2\}$, in the event space, $\{not \ \{1, 2\}\}$ is also a member of the space; to wit: $\{3, 4, 5, 6\}$. We see that the full sample space S is also a member of the event space. This allows us to talk about the event that "*something will occur*". The null set ensures that $\{not \ S\}$ would also be describable.

An example of an event space which is even smaller than the above example is the set $\{\{1, 3, 5\}, \{2, 4, 6\}, S, \phi\}$, which is appropriate to a situation in which one is interested only in the fact whether pips are even or odd. It is easy to see that this event space also has all the virtues that the earlier example has.

These properties, actually, sum up the qualities we want a general event space to possess. The set of desirable properties of a class of subsets of the sample space to qualify as an event space lead to the *formal definition* of the event space[12]:

If S is the sample space and E is a set of subsets of S such that

1. *the sample space S belongs to the set E,*

2. *if a subset A is a member of E, then its complement $S - A$ is also a member of E,*

---

[12]It is easy to see that for a discrete sample space, the power set fulfills all these properties for an event space. In fact, in this case, the power set provides the most *detailed* description of events for a given sample space.

3. *if the subsets A_i belong to E for $i = 1, 2, \ldots, n$, then their countable union $(\cup_{i=1}^{n} A_i)$ also belongs to E,*

then the set E qualifies as an event space[13].

It is not difficult to see why we have not listed the conditions that the event space should include all possible *intersections* and the *empty set*. It is because, these can be derived from the above properties[14].

Probability Measure

The first three properties of probability mentioned in the context of the frequentist interpretation in the previous section turn out to be an adequate characterization of *generalized probability* (with the first property slightly weakened to require non-negativity only). Technically, it is called a **probability measure**:

If S is a sample space, and E is an associated event space, then the probability P is a function $P : E \longrightarrow \mathbb{R}$ such that

1. *the function P is nonnegative: for all events A in E,*

$$P[A] \geq 0$$

2. *the **total probability** is unity:*

$$P[S] = 1$$

3. *the probability function is **countably additive**:*

$$P[\cup_{i=1}^{n} A_i] = \sum_{i=1}^{n} P[A_i]$$

where A_i are disjoint subsets of[15] E.

All the other properties that were listed for the frequentist definition of probability, follow from these properties.

We have, in this appendix, mentioned the frequentist interpretation, the classical interpretation and the Bayesian interpretation of probability. The

---

[13] In mathematical jargon, one says that the event space is required to include the sample space, and it must be *closed* under compliments and countable unions. Such a set (of subsets) is said to form a σ − algebra.

[14] One merely needs to use *De Morgan's* rules in conjunction with the *closure under complement* rule (the second property).

[15] As we have mentioned before, the requirement is actually pairwise disjointness, but we shall not bother about the subtle difference here.

defining properties of probability, listed above, are respected by all these different definitions.

For the frequentist definition, we have already listed the properties earlier. For the classical definition, the probability is again defined as a ratio, and it is extremely straightforward to prove that the above properties are obeyed. In the Bayesian interpretation, it is a requirement upon the *probability source* that generates the *degrees of belief* to produce probabilities that respect the above properties. Since the Bayesian updations use conditional probabilities as likelihoods, and yield conditional probabilities as posteriors, it is natural to ask whether the conditional probabilities (which constitute the key ingredient of the Bayesian methodology) have the desired properties of a probability measure. Indeed, it is not difficult to see that, given an event space, for every event A with probability $P(A) \neq 0$, one can define a function $\mathbb{P}(B)$ by

$$\mathbb{P}(B) = \frac{P(A \cap B)}{P(A)} \equiv P(B \mid A)$$

which satisfies all the required properties of a probability.

Appendix B

Linear Algebra

In this appendix, we fill in the proofs of many of the theorems that have been used in the text. Some theorems (without proofs) and some additional content important for quantum mechanics have also been included. In particular, linear functionals and dual spaces have been included to introduce the bra vector and the bra-ket notation. Although this has not been used in the text, this powerful language is standard in the quantum mechanics literature. Up until the point the bra-ket notation is introduced we have, however, used the traditional notation for vectors.

Vector Space

We shall denote the multiplication of two scalars, a and b, simply by juxtaposition ab, and multiplication of a vector $\vec{\alpha}$ by a scalar a by $a\vec{\alpha}$. The addition of vectors, $\vec{\alpha}$ and $\vec{\beta}$, and the addition of scalars, a and b, will both be denoted by the same symbol $+$ as $\vec{\alpha} + \vec{\beta}$ and $a + b$ respectively. It will always be clear from the context (i.e., the position of the operator with respect to the operands), which operation is intended. The additive identity in V will be denoted by $\vec{0}$.

Theorem: For all $\vec{\alpha} \in V$, $(-1)\vec{\alpha} = -\vec{\alpha}$.

Theorem: For all $a \in \mathbb{C}$ and $\vec{\alpha} \in V$, $a\vec{\alpha} = \vec{0}$ if and only if $a = 0$ or $\vec{\alpha} = \vec{0}$.

Theorem: If S is a subset of a vector space V, and if addition and scalar multiplication are closed in S, that is, if for all $\vec{\alpha}, \vec{\beta} \in S$, and for all $a \in F$, $\vec{\alpha} + \vec{\beta} \in S$ and $a\vec{\alpha} \in S$, then S is a subspace of V.

Theorem: The intersection of subspaces is a subspace.

It is easy to see that the union of two subspaces is not in general a subspace.

Theorem (Trivial Subspace): The subset $\{\vec{0}\}$ of a vector space comprising the additive identity $\vec{0}$ only, forms a subspace. This subspace is called the *trivial subspace*.

Linear independence, completeness and basis

Theorem (Replacement Theorem): If $B = \{\vec{\alpha_i} \, ; \, i = 1, 2, \ldots, n\}$ is a basis in V, and if $\vec{\beta} = \sum_{i=1}^{n} b_i \vec{\alpha_i}$ is some linear combination with scalars b_i such that $b_k \neq 0$ for some k, then the set

$$B' = \left\{ \vec{\alpha_1}, \vec{\alpha_2}, \ldots, \vec{\alpha}_{k-1}, \vec{\beta}, \vec{\alpha}_{k+1}, \ldots, \vec{\alpha_n} \right\}$$

is also a basis in V.

Proof (*completeness*): Let $\vec{\beta} = \sum_{i=1}^{n} b_i \vec{\alpha_i}$ where $b_k \neq 0$. Then (assuming all sums to run from 1 to n)

$$b_k \vec{\alpha}_k = \vec{\beta} - \sum_{j \neq k} b_j \vec{\alpha}_j$$

$$\text{or} \qquad \vec{\alpha}_k = b_k^{-1} \vec{\beta} - \sum_{j \neq k} b_k^{-1} b_j \vec{\alpha}_j$$

Now since B is a basis, every $\vec{\gamma} \in V$ can be written as a linear combination

$$\begin{aligned}
\vec{\gamma} &= \sum_i c_i \vec{\alpha}_i \\
&= \sum_{i \neq k} c_i \vec{\alpha}_i + c_k \vec{\alpha}_k \\
&= \sum_{i \neq k} c_i \vec{\alpha}_i + c_k \left(b_k^{-1} \vec{\beta} - \sum_{i \neq k} b_k^{-1} b_i \vec{\alpha}_i \right) \\
&= \sum_{i \neq k} \left(c_i - b_k^{-1} b_i \right) \vec{\alpha}_i + c_k b_k^{-1} \vec{\beta}
\end{aligned}$$

which proves the completeness of B'.

Proof (*linear independence*): Let us assume that (again with all sums running from 1 to n)

$$\sum_{i \neq k} e_i \vec{\alpha}_i + f \vec{\beta} = 0$$

$$\sum_{i \neq k} e_i \vec{\alpha}_i + f \left(\sum_{i=1}^{n} b_i \vec{\alpha_i} \right) = 0$$

$$\sum_{i \neq k} \left(e_i + f \, b_i \right) \vec{\alpha}_i + f \, b_k \vec{\alpha}_k = 0$$

Now by hypothesis $b_k \neq 0$. Hence from the linear independence of B, it follows that $f = 0$ and $e_i + f \, b_i = 0$ for $i \neq k$. But since $f = 0$ we have $e_i = 0$ for $i \neq k$. Therefore B' is a linearly independent set.

Theorem: If a set B is a basis in V, then every linearly independent set in V having the same number of elements as B will also constitute a basis.

Outline of proof: One merely needs to, iteratively, apply the replacement theorem to the given basis B, and the given set, say I, of linearly independent vectors until this set is exhausted, and all members of the basis B have been replaced by members of I. Let us call the basis after m replacements B^m. It may seem that the process mentioned above can fail if at some stage the only nonzero expansion coefficients of a vector belonging to $I - B^{k-1}$, when expanded in terms of the members of B^{k-1}, are the coefficients of the already replaced elements of B (i.e., members of $I \cap B^{k-1}$). In this case one can only replace an already replaced member of I, and not a hitherto unreplaced member of B. However, it is easy to convince oneself that this scenario will never actually arise owing to the linear independence of I.

Theorem: Every basis in a vector space has the same number of elements.

Outline of proof: This is a straightforward consequence of the previous theorem. Clearly, if we add an element into a set which is already a basis, it will cease to be independent, and if we exclude an element from a basis it will cease to be complete. Now if B_1 and B_2 are two bases such that one of them, say B_1, is larger than the other basis B_2, then one can replace B_2 by a proper subset B_1' of B_1 to form a new basis $B_2' = B_1'$. This implies B_1 is not independent (since members of $B_1 - B_1'$ can be expressed in terms of the members of B_1'). Thus we arrive at a contradiction.

This proves that the size of a basis is a characteristic of the vector space.

Theorem: Bases are the largest linearly independent sets in a vector space.

Outline of proof: Let B be a basis and let there exist a linearly independent set L larger than B. Any subset $L' \subset L$ having the same cardinality as B can completely replace B to form a basis. This means L can no longer be linearly independent leading to a contradiction.

Theorem: Bases are the smallest complete sets in a vector space.

Outline of proof: Let B be a basis and let there exist a complete set C smaller than B. There will always exist a subset $C' \subseteq C$ which will be complete and linearly independent so that it forms a basis[1]. Since C' must be smaller than B, this is impossible.

Representation

Definition (Homomorphisms and Isomorphisms of Vector Spaces): If $\{V, +, \star\}$ and $\{W, \oplus, \odot\}$ are two vector spaces, and $f : V \to W$ is a function

---

[1] If C is linearly independent $C' = C$ otherwise $C' \subset C$.

under which[2]

$$c_1 \vec{v}_1 + c_2 \vec{v}_2 \mapsto c_1 \vec{w}_1 + c_2 \vec{w}_2$$

where $\vec{w}_1 = f(\vec{v}_1)$ and $\vec{w}_2 = f(\vec{v}_2)$ for all $\vec{v}_1, \vec{v}_2 \in V$, and where c_1, c_2 are arbitrary scalars, then the function f is called a *homomorphism*. When the mapping f is one-to-one, it is called an *isomorphism*. The spaces V and W are referred to as being *homomorphic* or *isomorphic* to each other depending on whether f is a homomorphism or an isomorphism.

A homomorphism or isomorphism is, for obvious reason, called a *structure preserving map*[3].

Theorem: The vector space $\mathbb{C}^n$ is *isomorphic* to every complex vector space of dimension n.

Theorem (change of basis): If two n-tuples $[\beta] = \{b_i \, ; \, i = 1, 2, \ldots, n\}$ and $[\beta]' = \{b_i' \, ; \, i = 1, 2, \ldots, n\}$ are the representations of a vector $\vec{\beta} \in V$ with respect to two bases, $B = \{\vec{\alpha}_i \, ; \, i = 1, 2, \ldots, n\}$ and $B' = \{\vec{\alpha}_i' \, ; \, i = 1, 2, \ldots, n\}$, then the coordinates are related by

$$b_i' = \sum_{j=1}^{n} T_{ij} b_j$$

where the T_{ij} are given by

$$\vec{\alpha}_j = \sum_{i=1}^{n} T_{ij} \vec{\alpha}_i'$$

Clearly, the coefficients T_{ij} can be looked upon as the elements of a matrix T, so that the relations connecting the coordinates of $\vec{\beta}$ with respect to the bases B and B' can be written as a matrix relation

$$[\beta]' = T [\beta]$$

We say that the matrix T effects the transformation of coordinates under the change of basis B to B'. It is thus, often, referred to as the *transformation matrix*.

Theorem: A transformation matrix is necessarily invertible.

---

[2] For those unfamiliar, the symbol '$\mapsto$' means "maps to".

[3] The idea of a structure preserving map is a profoundly important concept, and it is not restricted to vector spaces. It is used on various algebraic structures and other places in mathematics.

Theorem: If $B = \{\vec{\alpha}_i; i = 1, 2, \ldots, n\}$ is a linearly independent set in V, and T is some $n \times n$ invertible matrix, then the set of vectors $B' = \{\vec{\alpha}'_i; i = 1, 2, \ldots, n\}$, where

$$\vec{\alpha}'_j = \sum_{i=1}^{n} T_{ij}^{-1} \vec{\alpha}_i$$

constitutes a linearly independent set in V.

Thus, if $B = \{\vec{\alpha}_i; i = 1, 2, \ldots, n\}$ is a basis in an n-dimensional vector space V, then every $n \times n$ invertible matrix T serves as a transformation matrix that effects a change of basis to a new basis $B' = \{\vec{\alpha}'_i; i = 1, 2, \ldots, n\}$, where the basis vectors $\vec{\alpha}'_i \in B'$ are given by the above equation.

Inner Product Space

Linear functionals and dual space

Definition (Linear Functional): A *linear functional* is a mapping f from a vector space V to the set of scalars $\mathbb{C}$:

$$f : V \longrightarrow \mathbb{C}$$

which obeys

$$f\left(\vec{\alpha} + \vec{\beta}\right) = f(\vec{\alpha}) + f\left(\vec{\beta}\right)$$
$$f(a\vec{\alpha}) = af(\vec{\alpha})$$

for all $\vec{\alpha}, \vec{\beta} \in V$ and for all $a \in \mathbb{C}$.

The above two properties imply that for all $\vec{\alpha}, \vec{\beta} \in V$ and for all $a, b \in \mathbb{C}$

$$f\left(a\vec{\alpha} + b\vec{\beta}\right) = af(\vec{\alpha}) + bf\left(\vec{\beta}\right)$$

that is, f depends linearly on its arguments. This gives the linear functional its name.

Theorem: Every vector $\vec{\alpha} \in V$ can be associated with a linear functional $f_{\vec{\alpha}}$ whose action on an arbitrary vector $\vec{\beta} \in V$ is defined to be equal to the inner product $\left(\vec{\alpha}, \vec{\beta}\right)$:

$$f_{\vec{\alpha}}\left(\vec{\beta}\right) = \left(\vec{\alpha}, \vec{\beta}\right)$$

Theorem (converse): Every linear functional $f\left(\vec{\beta}\right)$ defined on an inner product space V can be associated with a unique vector $\vec{\alpha} \in V$ such that

$$f\left(\vec{\beta}\right) = \left(\vec{\alpha}, \vec{\beta}\right)$$

where $\left(\vec{\alpha}, \vec{\beta}\right)$ is an appropriately defined inner product.

Bra-Ket Notation: So far, in this appendix, we have denoted vectors by Greek alphabets with arrows (e.g., $\vec{\alpha}$, $\vec{\beta}$, etc.), inner products by ordered pair of vectors (e.g., $\left(\vec{\alpha}, \vec{\beta}\right)$ etc.), and linear functionals corresponding to vectors by subscripted Latin alphabet (e.g., $f_{\vec{\alpha}}$, $f_{\vec{\beta}}$, etc.). We shall now move to a new notation: we shall denote a vector $\vec{\alpha}$ by $|\alpha\rangle$ and call it a **ket**. The linear functional corresponding to the ket $|\alpha\rangle$ will now be denoted by $\langle\alpha|$, and will be called a **bra**. Finally, the inner product $\left(\vec{\alpha}, \vec{\beta}\right)$, which by the theorem stated above, is equal to the action of the bra $\langle\alpha|$ on the ket $|\beta\rangle$ will be called the **braket** and denoted by $\langle\alpha|\beta\rangle$[4].

Definition (Addition and Scalar Multiplication on Bras): Let V^* be the set of all bras acting on V. We consider two functions,

$$\oplus \; : \; V^* \times V^* \; \longrightarrow \; V^*$$
$$\text{and} \quad \odot \; : \; \mathbb{C} \times V^* \; \longrightarrow \; V^*$$

called *addition of bras*, and *multiplication of a bra by a scalar* respectively. Their respective values will be denoted by $\langle\alpha|\oplus\langle\beta|$ and $c\odot\langle\alpha|$ for $\langle\alpha|, \langle\beta| \in V^*$ and $c \in F$. The functions $\oplus$ and $\odot$ are defined by

$$(\langle\alpha| \oplus \langle\beta|)\,|\gamma\rangle \;=\; \langle\alpha|\gamma\rangle + \langle\beta|\gamma\rangle$$
$$\text{and} \quad (c \odot \langle\alpha|)\,|\gamma\rangle \;=\; c\,\langle\alpha|\gamma\rangle$$

for all $|\gamma\rangle \in V$ and for all $c \in \mathbb{C}$.

We can use the simpler notations: $\langle\alpha| \oplus \langle\beta| \equiv \langle\alpha| + \langle\beta|$ and $c \odot \langle\alpha| \equiv c\,\langle\alpha|$. It will be clear from the operands which function is intended (i.e., whether the $+$ refers to addition of scalars, kets or bras).

Theorem (Dual Space): The space V^* of bras endowed with addition and multiplication by a scalar forms a vector space. The vector space V^* is called the *dual* of the vector space V.

The ket $|\alpha\rangle \in V$ and the corresponding bra $\langle\alpha| \in V^*$ are called *conjugates* of each other.

---

[4]This delightful piece of notation is due to P.A.M. Dirac.

A vector space V can also be looked upon as the dual of its dual space V^*. To see this we have to consider every ket $|\alpha\rangle \in V$ as a linear functional acting on the space of bras V^* such that the action of ket $|\alpha\rangle$ on a bra $\langle\beta|$ is defined to be equal to the inner product $\langle\beta|\alpha\rangle$.

Theorem: The bra corresponding to the ket $|\alpha\rangle + |\beta\rangle$ is given by $\langle\alpha| + \langle\beta|$.

Theorem: The bra corresponding to the ket $c\,|\alpha\rangle$ is given by $c^*\,\langle\alpha|$. Here c and c^* are complex numbers that are complex conjugates of each other.

The above theorems taken together are equivalent to the following statement: If $\{|\alpha_i\rangle\,;\, i = 1, 2, \ldots, n\}$ is a set of kets, then the bra corresponding to the linear combination of kets $\sum_{i=1}^{n} c_i\,|\alpha_i\rangle$ is given by $\sum_{i=1}^{n} c_i^*\,\langle\alpha_i|$.

Theorem (Dual Basis): If a set $B = \{|\alpha_i\rangle\,;\, i = 1, 2, \ldots, n\}$ forms a basis in V, then the set $B^* = \{\langle\alpha_i|\,;\, i = 1, 2, \ldots, n\}$ forms a basis in V^*. The basis B^* is called the *dual basis* of B.

Theorem (representation of bras in dual basis): If the set of vectors $B = \{|\alpha_i\rangle\,;\, i = 1, 2, \ldots, n\}$ be a basis in V, and $[\beta] \equiv (b_i\,;\, i = 1, 2, \ldots, n)$ be the representation of a ket $|\beta\rangle \in V$, then the representation of the corresponding bra $\langle\beta| \in V^*$ in the dual basis $B^* = \{\langle\alpha_i|\,;\, i = 1, 2, \ldots, n\}$ is $[\beta^*] \equiv (b_i^*\,;\, i = 1, 2, \ldots, n)$.

It is customary to write the representations of the bra vectors as row matrices. To indicate a bra vector in a representation, we shall use a superscript '$*$' on the alphabet denoting the vector and enclose it in a square bracket as usual.

Theorem (change of dual basis): If T is a transformation matrix that effects the transformation of coordinates of kets in V from a basis B to B', then the transformation matrix that effects the transformation of coordinates of the bras in V^* from the dual basis B^* to B'^* is the Hermitian conjugate $T^\dagger$ of the matrix T:

$$[\beta^*]' = [\beta^*]\,T^\dagger$$

for all $\langle\beta| \in V^*$.

Orthonormal bases

Theorem (Gram-Schmidt Orthogonalization): If B is a linearly independent set of n kets, then it is always possible to construct an orthonormal set of n kets comprising linear combinations of the elements of B.

Proof: We prove this theorem by explicit construction. The key idea underlying the construction can be easily visualized. If one subtracts from a vector, the resultant of its projections along a set of m orthonormal vectors, then the new vector is orthogonal to each of the vectors of the orthonormal set. Normalizing this vector then furnishes a set of $m+1$ orthonormal vectors.

Let $B = \{|\alpha_i\rangle\,;\, i = 1, 2, \ldots, n\}$ be a linearly independent set. We construct a set $B' = \{|\beta_i\rangle\,;\, i = 1, 2, \ldots, n\}$ as follows:

$$|\beta_1\rangle = \frac{|\alpha_1\rangle}{\|\alpha_1\|}$$

$$|\gamma_2\rangle = |\alpha_2\rangle - |\beta_1\rangle\langle\beta_1|\alpha_2\rangle$$

$$|\beta_2\rangle = \frac{|\gamma_2\rangle}{\|\gamma_2\|}$$

$$|\gamma_3\rangle = |\alpha_3\rangle - |\beta_1\rangle\langle\beta_1|\alpha_3\rangle - |\beta_2\rangle\langle\beta_2|\alpha_3\rangle$$

$$|\beta_3\rangle = \frac{|\gamma_3\rangle}{\|\gamma_3\|}$$

$$\vdots$$

$$|\gamma_n\rangle = |\alpha_n\rangle - \sum_{i=1}^{n-1}|\beta_i\rangle\langle\beta_i|\alpha_n\rangle$$

$$|\beta_n\rangle = \frac{|\gamma_n\rangle}{\|\gamma_n\|}$$

It is easy to check explicitly that the set B' is an orthonormal set.

It immediately follows that, given a set of any n independent vectors in an n-dimensional inner product space, it is always possible to construct an orthonormal basis out of linear combinations of the n independent vectors.

Theorem (representation of bras in orthonormal basis): If $B = \{|\alpha_i\rangle\,;\, i = 1, 2, \ldots, n\}$ is an orthonormal basis in V, then the dual basis B^* is also orthonormal in the dual space V^*.

The coordinates b_i^* of the bra $\langle\beta|$ in the dual basis B^* are given by

$$b_i^* = \langle\beta|\alpha_i\rangle$$

Linear Operators

Linear operators on dual space

Definition (Linear Operator on Dual Space): Let $\hat{A}$ be a linear operator acting on an inner product space V. Then we can define a *linear operator on* *the dual space V^* (which we can choose to denote by the same notation $\hat{A}$)* whose action on an arbitrary bra vector $\langle\alpha| \in V^*$ is specified by the condition

$$\left(\langle\alpha|\,\hat{A}\right)|\beta\rangle = \langle\alpha|\left(\hat{A}\,|\beta\rangle\right)$$

for all $|\beta\rangle \in V$.

The definition makes the position of the parenthesis irrelevant so that we can drop it altogether, and unambiguously use the notation $\left\langle\alpha|\hat{A}|\beta\right\rangle$.

A linear operator defined on an inner product space naturally induces a linear operator acting on the dual space.

Theorem: The action of the induced operator on the bra vectors is linear:

$$\left(a\,\langle\alpha| + b\,\langle\beta|\right)\hat{A} = a\left(\langle\alpha|\,\hat{A}\right) + b\left(\langle\beta|\,\hat{A}\right)$$

for all $\langle\alpha|\,,\langle\beta| \in V$ and all $a,b \in \mathbb{C}$.

Hermitian operator

Theorem: If $\hat{A}$ be a linear operator acting on an inner product space V such that

$$\hat{A}\,|\alpha\rangle = |\alpha\rangle'$$

where $|\alpha\rangle\,,|\alpha\rangle' \in V$, then a necessary and sufficient condition that an operator $\hat{A}^\dagger$ be the Hermitian conjugate of $\hat{A}$ is

$$\langle\alpha|\,\hat{A}^\dagger = \langle\alpha|'$$

where $\langle\alpha|$ and $\langle\alpha|'$ are the bra vectors conjugate to the kets $|\alpha\rangle$ and $|\alpha\rangle'$.

This can be used as an equivalent definition of the Hermitian conjugate (or adjoint) operator.

Theorem: A linear operator acting on an inner product space V is Hermitian if and only if $\left\langle\alpha|\hat{A}|\alpha\right\rangle$ is real for all $|\alpha\rangle \in V$.

Theorem: Any linear operator can be written as a sum of a Hermitian operator and an anti-Hermitian operator.

Proof: Let $\hat{A}$ be a linear operator. Then we can write

$$\hat{A} = \left(\frac{\hat{A} + \hat{A}^\dagger}{2}\right) + \left(\frac{\hat{A} - \hat{A}^\dagger}{2}\right)$$

Evidently, the operator $(\hat{A}+\hat{A}^\dagger)/2$ is Hermitian and $(\hat{A}-\hat{A}^\dagger)/2$ is anti-Hermitian.

Unitary operator

For the next two theorems we will use the following notation for inner product: $\langle \beta | \alpha \rangle \equiv (|\beta\rangle, |\alpha\rangle)$.

Theorem: A linear operator $\hat{U}$ acting on an inner product space V is unitary if and only if the inner product is invariant under the action of $\hat{U}$:

$$\left(\hat{U} |\alpha\rangle, \hat{U} |\beta\rangle \right) = (|\alpha\rangle, |\beta\rangle)$$

for all $|\alpha\rangle, |\beta\rangle \in V$.

There is actually a stronger theorem that furnishes an equivalent definition of unitarity:

Theorem: A linear operator $\hat{U}$ acting on an inner product space V is unitary if and only if the norm of $|\alpha\rangle$ is equal to the norm of $\hat{U} |\alpha\rangle$:

$$\left(\hat{U} |\alpha\rangle, \hat{U} |\alpha\rangle \right) = (|\alpha\rangle, |\alpha\rangle)$$

for all $|\alpha\rangle \in V$.

Representation of linear operators

Theorem (representation of a linear operator acting on V): Let $B = \{|\alpha_i\rangle \,;\, i = 1, 2, \ldots, n\}$ be a basis in the vector space V. If $\hat{A}$ be a linear operator acting on V such that $\hat{A} |\beta\rangle = |\beta\rangle'$ where $|\beta\rangle, |\beta\rangle' \in V$, then the coordinates b_i and b_i' of the representations of the vectors $|\beta\rangle$ and $|\beta\rangle'$ in the basis B are related by

$$b_i' = \sum_{j=1}^{n} A_{ij} b_j$$

where A_{ij} is given by

$$\hat{A} |\alpha_j\rangle = \sum_{i=1}^{n} A_{ij} |\alpha_i\rangle$$

The set of elements $\{A_{ij} \,;\, i, j = 1, 2, \ldots, n\}$ is called the *representation of the linear operator* $\hat{A}$ in the basis B.

If B is an orthonormal basis, then

$$A_{ij} = \langle \alpha_i | \hat{A} |\alpha_j\rangle$$

The elements A_{ij} can be written out as a $n \times n$ square matrix A so that the representations $[\beta]$ and $[\beta']$ of $|\beta\rangle$ and $|\beta\rangle'$ are related by the matrix equation

$$[\beta'] = A \, [\beta]$$

Theorem (representation of a linear operator acting on V^*): If a linear operator $\hat{A}$ defined on some vector space V induces a linear operator $\hat{A}$ acting on the dual space V^* such that $\langle\beta|\,\hat{A} = \langle\beta|'$ for $\langle\beta|\,,\langle\beta|' \in V^*$, then the coordinates b_i^* and $b_i^{*\prime}$ of the representations of the bra vectors $\langle\beta|$ and $\langle\beta|'$, conjugate to the kets $|\beta\rangle$ and $|\beta\rangle'$, in the dual basis B^* are related by

$$b_i^{*\prime} = \sum_{j=1}^{n} A_{ji}b_j^*$$

Thus, the representations $[\beta^*]$ and $[\beta^{*\prime}]$ of $\langle\beta|$ and $\langle\beta|'$ are related by the matrix equation

$$[\beta^{*\prime}] = [\beta^*]\,A$$

Here, recall that the representations of bra vectors are written as row matrices.

Thus, the representation of a linear operator acts on the representations of vectors (both, kets and bras) by matrix multiplication. Moreover, a linear operator acting on a vector space and the corresponding linear operator acting on the dual space are represented, respectively, with respect to a basis and its dual, by the same matrix.

Theorem (change of basis): The representations A and A' of a linear operator $\hat{A}$ with respect to two bases B and B' are related by

$$A' = T A T^{-1}$$

where T is the transformation matrix that effects the transformation from B to B'.

If the bases B and B' are orthonormal, then $T^{-1} = T^\dagger$, so that

$$A' = T A T^\dagger$$

Eigen Systems

Determination of eigenvectors

Consider the eigenvalue equation of the operator $\hat{A}$, for some eigenvalue λ, represented by the matrix A in some basis $B = \{|\alpha_i\rangle\,;\, i = 1, 2, \ldots, n\}$:

$$\sum_{j=1}^{n} A_{ij}\beta_j = \lambda\beta_i$$

where β_j are components of the eigenvector in the basis B. The characteristic equation implies that not all the equations are independent[5]. Let the number of linearly independent equations constituting the eigenvalue equation be $n-p$. Thus, there will be infinitely many solutions for the variables β_i. However, we can determine in terms of the p variables $\{\beta_i \, ; \, i = 1, 2, \ldots, p\}$ say, the remaining $n - p$ variables $\{\beta_i \, ; \, i = p+1, p+2, \ldots, n\}$ as linear combinations, so that for β_i with $p < i \le n$, we can write $\beta_i = \sum_{j=1}^{p} C_{ij}\beta_j$ for some choice of scalars[6] C_{ij}. For β_i with $1 \le i \le p$, one can trivially write a similar equation with $C_{ij} = \delta_{ij}$. Now an arbitrary eigenvector can be written as

$$
\begin{aligned}
|\beta\rangle &= \sum_{i=1}^{n} \beta_i \, |\alpha_i\rangle \\
&= \sum_{i=1}^{n} \left(\sum_{j=1}^{p} C_{ij}\beta_j \right) |\alpha_i\rangle \\
&= \sum_{j=1}^{p} \beta_j \left(\sum_{i=1}^{n} C_{ij} \, |\alpha_i\rangle \right) \\
&= \sum_{j=1}^{p} \beta_j \, |\gamma_j\rangle
\end{aligned}
$$

where the coefficients β_j are arbitrary and $|\gamma_j\rangle = \sum_{i=1}^{n} C_{ij} \, |\alpha_i\rangle$. Thus every eigenvector is a linear combination of the p linearly independent eigenvectors $|\gamma_j\rangle$ with j running from[7] 1 to p. The linear independence of vectors $|\gamma_j\rangle$ can be easily seen from the condition $C_{ij} = \delta_{ij}$ when $i = 1, 2, \ldots, p$.

If $n - p$ is the number of linearly independent equations among the n linear homogeneous equations comprising the eigenvalue equation, then p is degree of degeneracy or the geometric multiplicity of the eigenvalue.

Theorem: If an eigenvalue is a root of the characteristic equation of multiplicity one, then it is nondegenerate (i.e., its eigensubspace has dimension one). If an eigenvalue is a root of the characteristic equation of multiplicity $h > 1$, then the degree of degeneracy p is restricted to $1 \le p \le h$ (i.e., the dimension of the associated eigensubspace is at least one and at most h). In other words,

the geometric multiplicity of an eigenvalue is at least one and at most equal to its algebraic multiplicity.

---

[5]The determinant of a matrix is zero if and only if the rows (or columns) are linearly dependent.

[6]To see this, just use Cramer's rule.

[7]That the vectors $|\gamma_j\rangle$ are eigenvectors can be seen by observing that the β_js are arbitrary. Hence, one may choose $\beta_j = \delta_{jk}$ with k taking values from 1 to p.

Outline of proof: By using the replacement theorem, one can construct a basis B' by replacing p of the elements of the initial basis B (in which the eigenvalue equation was cast) by the p independent eigenvectors of some eigenvalue λ, and re-arrange to keep these as the first p elements of the new basis. The representation A' of the operator $\hat{A}$ (in question) in the new basis B' will consist of four blocks $A_{(1,1)}, A_{(1,2)}, A_{(2,1)}$ and $A_{(2,2)}$:

$$A' = \begin{bmatrix} A_{(1,1)} & A_{(1,2)} \\ A_{(2,1)} & A_{(2,2)} \end{bmatrix}$$

where $A_{(1,1)} = \lambda I_{p \times p}$, $A_{(2,1)}$ is a $(n-p) \times p$ dimensional null matrix and $A_{(1,2)}$, $A_{(2,2)}$ are $p \times (n-p)$ and $(n-p) \times (n-p)$ dimensional matrices respectively[8]. The characteristic equation when cast in terms of the basis B' (using μ as the undetermined eigenvalue) will be:

$$\det \left[A' - \mu I_{n \times n} \right] = 0$$
$$\det \left[\lambda I_{p \times p} - \mu I_{p \times p} \right] \det \left[A_{(2,2)} - \mu I_{(n-p) \times (n-p)} \right] = 0$$
$$(\lambda - \mu)^p \det \left[A_{(2,2)} - \mu I_{(n-p) \times (n-p)} \right] = 0$$

so that the multiplicity of the root λ of μ is at least p.

Diagonalization

Theorem: If $\hat{A}$ is a diagonalizable linear operator acting on a vector space V represented by a matrix A in some basis $B = \{|\alpha_i\rangle ; i = 1, 2, \ldots, n\}$, and $B_A = \{|\beta_j\rangle ; j = 1, 2, \ldots, n\}$ is a basis consisting of eigenvectors of $\hat{A}$, then the diagonal representation A_d of $\hat{A}$ will be given by

$$A_d = T A T^{-1}$$

where T is the transformation matrix given by

$$|\beta_j\rangle = \sum_{i=1}^{n} T_{ij}^{-1} |\alpha_i\rangle$$

i.e., the columns of the inverse transformation matrix T^{-1} are the eigenvectors of $\hat{A}$ represented in the basis B.

Diagonalizability of Hermitian and unitary operators

Theorem: Eigenvalues of a Hermitian operator are real.

---

[8] Here $I_{q \times q}$ is, obviously, the q dimensional identity matrix.

Proof: If $\hat{H}$ is a Hermitian operator having an eigenvector $|\psi\rangle$ with eigenvalue λ, we have

$$\left\langle \psi|\hat{H}|\psi \right\rangle = \lambda \left\langle \psi|\psi \right\rangle$$

$$\left\langle \psi|\hat{H}|\psi \right\rangle = \lambda^* \left\langle \psi|\psi \right\rangle$$

where the second equation was obtained by taking the Hermitian conjugation of the first equation and using the Hermiticity of $\hat{H}$. The result ($\lambda = \lambda^*$) follows by taking the difference.

Theorem: Eigenvectors of a Hermitian operator corresponding to distinct eigenvalues are orthogonal.

Proof: If $\hat{H}$ is a Hermitian operator, then for two arbitrary eigenvectors $|\phi\rangle$ and $|\psi\rangle$ with distinct eigenvalues λ and μ, we have

$$\left\langle \phi|\hat{H}|\psi \right\rangle = \lambda \left\langle \phi|\psi \right\rangle$$

$$\left\langle \psi|\hat{H}|\phi \right\rangle = \mu \left\langle \psi|\phi \right\rangle$$

Taking the Hermitian conjugation of the second equation, using the Hermiticity of $\hat{H}$, and using the fact that the eigenvalues (in particular, μ) is real, we get

$$\left\langle \phi|\hat{H}|\psi \right\rangle = \mu \left\langle \phi|\psi \right\rangle$$

Since $\lambda \neq \mu$, the result ($\langle\phi|\psi\rangle = 0$) follows by taking the difference of this equation from the first.

Theorem: It is always possible to construct an orthonormal basis out of the eigenvectors of a Hermitian operator.

A Hermitian matrix (the representation of some Hermitian operator in an orthonormal basis) is thus always diagonalizable by a unitary transformation.

Outline of proof: By the fundamental theorem of algebra, there will exist at least one eigenvalue α and one eigenvector $|\alpha\rangle$ of the operator $\hat{A}$ in question. We can always construct an orthonormal basis B containing $|\alpha\rangle$. Owing to the Hermiticity of $\hat{A}$, the linear span V' of $B - \{|\alpha\rangle\}$ (the set constructed by excluding $|\alpha\rangle$ from B) will be an *invariant subspace* of $\hat{A}$ (i.e., $\hat{A}|\beta\rangle \in V'$ whenever $|\beta\rangle \in V'$)[9] . We can then repeat the argument on V'. The result will follow from finite induction.

Theorem (converse): If the eigenvalues of a linear operator are all real, and if an orthogonal basis can be constructed out of the eigenvectors of this operator, then the operator is Hermitian.

---

[9]V' is called the orthogonal complement of $|\alpha\rangle$.

Outline of proof: Let us, for this proof, write the inner product $\langle \alpha | \beta \rangle$ as $(|\alpha\rangle, |\beta\rangle)$. If $\{|\alpha_i\rangle; i = 1, 2, \ldots, n\}$ is an orthonormal basis of eigenvectors of an operator $\hat{A}$ with α_i being the eigenvalue associated with $|\alpha_i\rangle$, then for two arbitrary vectors $|\phi\rangle = \sum_{i=1}^{n} \phi_i |\alpha_i\rangle$ and $|\psi\rangle = \sum_{i=1}^{n} \psi_i |\alpha_i\rangle$, a direct evaluation gives[10] $\left(\hat{A} |\phi\rangle, |\psi\rangle \right) = \sum_{i=1}^{n} \phi_i^* \psi_i \alpha_i = \left(|\phi\rangle, \hat{A} |\psi\rangle \right)$.

Theorem[11]**:** Eigenvalues of unitary operators are unimodular (i.e., they have absolute value one).

Proof: Here we denote the inner product $\langle \alpha | \beta \rangle$ also by $(|\alpha\rangle, |\beta\rangle)$. If $\hat{U}$ is an operator having an eigenvector $|\psi\rangle$ with eigenvalue λ, we have

$$\left(\hat{U} |\psi\rangle, \hat{U} |\psi\rangle \right) = \lambda^* \lambda \langle \psi | \psi \rangle$$

But if $\hat{U}$ is unitary

$$\left(\hat{U} |\psi\rangle, \hat{U} |\psi\rangle \right) = \left\langle \psi | \hat{U}^\dagger \hat{U} | \psi \right\rangle = \langle \psi | \psi \rangle$$

whence the result follows by taking the difference.

Theorem: Eigenvectors of a unitary operator corresponding to distinct eigenvalues are orthogonal.

Proof: If $\hat{U}$ is a unitary operator having an eigenvector $|\psi\rangle$ with eigenvalue λ, then $|\psi\rangle$, clearly, is an eigenvector of $\hat{U}^\dagger$ with eigenvalue $1/\lambda$. Then, if $\hat{U}$ has two eigenvectors $|\phi\rangle$ and $|\psi\rangle$ with distinct eigenvalues λ and μ, we have

$$\left\langle \psi | \hat{U} | \phi \right\rangle = \mu \langle \psi | \phi \rangle$$

$$\left\langle \phi | \hat{U}^\dagger | \psi \right\rangle = \frac{1}{\lambda} \langle \phi | \psi \rangle$$

Taking the complex conjugate of the second equation

$$\left\langle \psi | \hat{U} | \phi \right\rangle = \frac{1}{\lambda^*} \langle \psi | \phi \rangle$$

Taking the difference with the first equation and using the unimodularity of the eigenvalues, the result follows.

Theorem: It is always possible to construct an orthonormal basis out of the eigenvectors of a unitary operator.

---

[10]The apparent restriction to nondegenerate eigenvalues is insignificant for the argument. Indeed one may simply allow the eigenvalues α_i to be "not necessarily distinct" (although the associated kets $|\alpha_i\rangle$ are distinct).

[11]The proofs of this and following three theorems are similar to that for the corresponding theorems for Hermitian operators.

A unitary matrix (representation of some unitary operator in an orthonormal basis) is thus always diagonalizable by a unitary transformation.

Outline of proof: Observing that an unitary operator and its Hermitian conjugate share the same eigenvectors, the proof follows exactly along the same lines as the proof of the corresponding theorem for the Hermitian operator.

Theorem (converse): If the eigenvalues of a linear operator are all unimodular, and if an orthogonal basis can be constructed out of the eigenvectors of this operator, then the operator is unitary.

Outline of proof: If $\{|\alpha_i\rangle \, ; \, i = 1, 2, \ldots, n\}$ is an orthonormal basis of eigenvectors of an operator $\hat{A}$ with α_i being the eigenvalue associated with $|\alpha_i\rangle$, then for an arbitrary vector $|\psi\rangle = \sum_{i=1}^{n} \psi_i |\alpha_i\rangle$, a direct evaluation gives

$$\left\langle \psi | \hat{A}^\dagger \hat{A} | \psi \right\rangle = \left(\hat{A} |\psi\rangle, \hat{A} |\psi\rangle \right) = \sum_{i=1}^{n} \psi_i^* \psi_i = \langle \psi | \psi \rangle$$

where in the first equality we have written the inner product using $\langle \alpha | \beta \rangle \equiv (|\alpha\rangle, |\beta\rangle)$ on the right hand side.

Direct Sum of Subspaces[12]

Definition (Direct Sum of Subspaces): If H and K are two subspaces of a vector space V such that every element $|\alpha\rangle \in V$ can be *uniquely* written as $|\alpha\rangle = |\beta\rangle + |\gamma\rangle$ where $|\beta\rangle \in H$ and $|\gamma\rangle \in K$, then V is said to be a *direct sum* of H and K.

We denote the direct sum by $V = H \oplus K$.

---

[12]The concept of a direct sum can be introduced without resorting to subspaces: A direct sum of the vector spaces H and K is the set $V = H \times K$ on which two functions, addition and multiplication by a scalar, are defined as

- $\left(\vec{h}_1, \vec{k}_1 \right) + \left(\vec{h}_2, \vec{k}_2 \right) = \left(\vec{h}_1 + \vec{h}_2, \vec{k}_1 + \vec{k}_2 \right)$
- $a \left(\vec{h}_1, \vec{k}_1 \right) = \left(a\vec{h}_1, a\vec{k}_1 \right)$

for all $\vec{h}_1, \vec{h}_2 \in H$, $\vec{k}_1, \vec{k}_2 \in K$ and $a \in \mathbb{C}$. It is easy to check that V forms a vector space.

Now the set of vectors of the form $\left(\vec{h}, \vec{0} \right)$, where $\vec{h} \in H$ and $\vec{0}$ is the additive identity in K, forms a subspace of V that is isomorphic to H. Similarly, the set of vectors of the form $\left(\vec{0}, \vec{k} \right)$, where $\vec{0}$ is now the additive identity in H and $\vec{k} \in K$, forms a subspace that is isomorphic to K. The space V is the direct sum of these subspaces according to the definition provided in the text. This establishes the connection between the definitions.

Theorem: A necessary and sufficient condition for $V = H \oplus K$ is that $H \cap K = \{|\,\rangle\}$ (the trivial subspace), and every element $|\alpha\rangle \in V$ can be written as $|\alpha\rangle = |\beta\rangle + |\gamma\rangle$ where $|\beta\rangle \in H$ and $|\gamma\rangle \in K$.

Theorem: If V is a direct sum of two subspaces, H and K, and $B_H = \{|h_i\rangle\,;\, i = 1, 2, \ldots, n_H\}$ and $B_K = \{|k_i\rangle\,;\, i = 1, 2, \ldots, n_K\}$ are bases in the subspaces H and K respectively, then the union of the bases $B = B_H \cup B_K$ is a basis in V.

Conversely, if $B = B_H \cup B_K$ is a basis in V, where $B_H \cap B_K = \phi$ (null set), then V is a direct sum of the linear spans of B_H and B_K.

Projection operator

Definition (Projection Operator): If $V = H \oplus K$ so that for all $|\alpha\rangle \in V$ we can write $|\alpha\rangle = |\alpha_H\rangle + |\alpha_K\rangle$ where $|\alpha_H\rangle \in H$ and $|\alpha_K\rangle \in K$, then we define the *projection operator* $\hat{P}_H$ onto the subspace H *along* K by[13]

$$\hat{P}_H |\alpha\rangle = |\alpha_H\rangle$$

Note that this requires $\hat{P}_H |\alpha_K\rangle = |\,\rangle$, where $|\,\rangle$ is the additive identity vector in V. Corresponding to a given H, there could be several choices (actually infinite) of K such that $V = H \oplus K$. Therefore, the specification of the projector onto H depends on the choice of K, and we say "projection of a vector onto H *along* K".

Theorem: The projection operator is linear.

Theorem: A necessary and sufficient condition for a linear operator to be a projection operator is that it is *idempotent*:

$$\hat{P}^2 = \hat{P}$$

Proof (*necessity*): Let $\hat{P}$ be a projector onto a subspace H along K of a vector space $V = H \oplus K$. For any $|\alpha\rangle \in V$ we then have $\hat{P}|\alpha\rangle \in H$. Now for any $|\beta\rangle \in H$, the sum $|\beta\rangle = |\beta\rangle + |\,\rangle$ is a unique expansion of $|\beta\rangle$ (as a sum of two vectors, one from H and the other from K). So we must have $\hat{P}|\beta\rangle = |\beta\rangle$. Taking $\hat{P}|\alpha\rangle = |\beta\rangle$, we have $\hat{P}\hat{P}|\alpha\rangle = \hat{P}|\alpha\rangle$.

Proof (*sufficiency*): Let $\hat{P}$ be an operator acting on a vector space V satisfying $\hat{P}^2 = \hat{P}$. Now for any vector $|\alpha\rangle \in V$, one can always write, $|\alpha\rangle = \hat{P}|\alpha\rangle + \left(\hat{I} - \hat{P}\right)|\alpha\rangle$ for any operator $\hat{P}$. Here $\hat{I}$ is the identity operator. We observe that $\hat{P}\hat{P}|\alpha\rangle = \hat{P}|\alpha\rangle$ and $\hat{P}\left(\hat{I} - \hat{P}\right)|\alpha\rangle = |\,\rangle$. If we define

---

[13]Sometimes the projection operator is simply called the projector.

subspaces H and K by the conditions that $H = \left\{ |\beta\rangle \,:\, \hat{P}\,|\beta\rangle = |\beta\rangle \right\}$ and $K = \left\{ |\beta\rangle \,:\, \hat{P}\,|\beta\rangle = |\,\rangle \right\}$, then $\hat{P}\,|\alpha\rangle \in H$ and[14] $\left(\hat{I} - \hat{P} \right) |\alpha\rangle \in K$. Moreover, $|\beta\rangle \in H$ and $|\beta\rangle \in K$ implies $|\beta\rangle = \hat{P}\,|\beta\rangle = |\,\rangle$. Thus $H \cap K = \{|\,\rangle\}$ (trivial subspace). Hence $V = H \oplus K$, and $\hat{P}$ is a projector onto H along K.

Theorem: The eigenvalues of projection operators are either 0 or 1.

Theorem: A necessary and sufficient condition for a linear operator $\hat{P}_H$ to be a projector is that $\hat{I} - \hat{P}_H$ is also a projector, where $\hat{I}$ is the identity operator.

Definition (Orthogonal Projector): If $V = H \oplus H^\perp$ where $H^\perp$ is a subspace such that every vector in $H^\perp$ is orthogonal to every vector in H, then the associated projectors are called *orthogonal projectors*. The subspace $H^\perp$ is uniquely determined by H, and hence, one refers to the *projector onto H along $H^\perp$* simply as the *projector onto H*.

Theorem: A necessary and sufficient condition for an operator $\hat{P}$ to be an orthogonal projector is

$$\hat{P} = \hat{P}^2 = \hat{P}^\dagger$$

Thus, orthogonal projection operators are Hermitian.

Theorem: A Hermitian operator $\hat{H}$ with m eigenvalues can be written as

$$\hat{H} = \sum_{i=1}^{m} a_i \hat{P}_i$$

where a_i are the m real eigenvalues of $\hat{H}$, and $\hat{P}_i$ are the orthogonal projectors onto the eigensubspaces associated with the eigenvalues a_i. The projectors $\hat{P}_i$ obey

$$\hat{P}_i \hat{P}_j = \hat{0} \qquad \text{for } i \neq j$$
$$\text{and} \qquad \sum_{i=1}^{m} \hat{P}_i = \hat{I}$$

Here $\hat{0}$ is the *null operator* defined by $\hat{0}\,|\psi\rangle = |\,\rangle$ for all $|\psi\rangle$ in the vector space on which $\hat{H}$ acts, and the operator $\hat{I}$ is the identity operator acting on that space. The symbol $|\,\rangle$, as always, refers to the additive identity vector.

Definition (Outer Product): Let Θ be the set of linear operators acting on V. We construct a function

$$\rangle\langle : V \times V^* \longrightarrow \Theta$$

which we call *outer product*, and whose value we denote by $|\alpha\rangle\langle\beta|$ for all $|\alpha\rangle \in V$ and all $\langle\beta| \in V^*$. The operator $|\alpha\rangle\langle\beta|$ is defined by

$$(|\alpha\rangle\langle\beta|)\,|\gamma\rangle = |\alpha\rangle\,\langle\beta|\gamma\rangle$$

for all $|\gamma\rangle \in V$.

---

[14]It is only too easy to check that H and K are subspaces.

Theorem: If $V = H \oplus H^{\perp}$, and $B_H = \{|h_i\rangle ; i = 1, 2, \ldots, n_H\}$ and $B_{H^{\perp}} = \{|k_i\rangle ; i = 1, 2, \ldots, n_{H^{\perp}}\}$ are orthonormal bases in H and $H^{\perp}$ respectively, then the linear operators

$$\hat{P}_H = \sum_{i=1}^{n_H} |h_i\rangle \langle h_i|$$

$$\text{and} \quad \hat{P}_{H^{\perp}} = \sum_{i=1}^{n_{H^{\perp}}} |k_i\rangle \langle k_i|$$

are orthogonal projectors onto the subspaces H and $H^{\perp}$. Further

$$\hat{P}_{H^{\perp}} = \hat{I} - \hat{P}_H$$

where $\hat{I}$ is the identity operator.

Tensor Product Space

In what follows, if there is no risk of ambiguity, we shall often use the notation $B^1 \otimes B^2$ to denote a set constructed out of all possible tensor products of two sets of vectors B^1 and B^2, where $B^1 \subseteq V^1$ and $B^2 \subseteq V^2$.

Theorem: If V^1, V^2 and V are vector spaces and $\otimes$ is a bilinear function

$$\otimes : V^1 \times V^2 \longrightarrow V$$

such that we have one basis $B = B^1 \otimes B^2$ where B^1 and B^2 are bases in V^1 and V^2 respectively, then for every other pair of bases C^1 and C^2 belonging respectively to V^1 and V^2, the set $C = C^1 \otimes C^2$ will also form a basis in[15] V.

Proof: Let $\left|\beta_i^{(1)}\right\rangle$ and $\left|\beta_j^{(2)}\right\rangle$ denote the basis vectors in B^1 and B^2 respectively, and similarly, let $\left|\gamma_k^{(1)}\right\rangle$ and $\left|\gamma_l^{(2)}\right\rangle$ denote the respective basis vectors in C^1 and C^2. We assume the dimensions of V^1 and V^2 to be d_1 and d_2 respectively. We have to prove the completeness and linear independence of the set $C = \left\{\left|\gamma_k^{(1)}\right\rangle \otimes \left|\gamma_l^{(2)}\right\rangle\right\} ; k = 1, 2, \ldots, d_1; l = 1, 2, \ldots, d_2$.

---

[15]Thus, it is not necessary to test the second defining property of a tensor product space (defined in chapter 9) for all possible bases of V^1 and V^2. It is enough that one basis such as $B = B^1 \otimes B^2$ can be found in V.

Completeness: If $|\psi\rangle \in V$, then it can be expanded as

$$|\psi\rangle = \sum_{i=1}^{d_1}\sum_{j=1}^{d_2} \psi_{ij} \left|\beta_i^{(1)}\right\rangle \otimes \left|\beta_j^{(2)}\right\rangle$$

$$= \sum_{i=1}^{d_1}\sum_{j=1}^{d_2} \psi_{ij} \left(\sum_{k=1}^{d_1} b_{ki}^1 \left|\gamma_k^{(1)}\right\rangle\right) \otimes \left(\sum_{l=1}^{d_2} b_{lj}^2 \left|\gamma_l^{(2)}\right\rangle\right)$$

$$= \sum_{k=1}^{d_1}\sum_{l=1}^{d_2} \left(\sum_{i=1}^{d_1}\sum_{j=1}^{d_2} b_{ki}^1 b_{lj}^2 \psi_{ij}\right) \left|\gamma_k^{(1)}\right\rangle \otimes \left|\gamma_l^{(2)}\right\rangle$$

where $\left|\beta_i^{(1)}\right\rangle = \sum_{k=1}^{d_1} b_{ki}^1 \left|\gamma_k^{(1)}\right\rangle$, $\left|\beta_j^{(2)}\right\rangle = \sum_{l=1}^{d_2} b_{lj}^2 \left|\gamma_l^{(2)}\right\rangle$ and ψ_{ij}, b_{ki}^1 and b_{lj}^2 denote the appropriate expansion coefficients.

Linear independence: Let h_{kl} be a set of coefficients such that

$$\sum_{k=1}^{d_1}\sum_{l=1}^{d_2} h_{kl} \left|\gamma_k^{(1)}\right\rangle \otimes \left|\gamma_l^{(2)}\right\rangle = 0$$

$$\sum_{k=1}^{d_1}\sum_{l=1}^{d_2} h_{kl} \left(\sum_{i=1}^{d_1} c_{ik}^1 \left|\beta_i^{(1)}\right\rangle\right) \otimes \left(\sum_{j=1}^{d_2} c_{jl}^2 \left|\beta_j^{(2)}\right\rangle\right) = 0$$

$$\sum_{i=1}^{d_1}\sum_{j=1}^{d_2} \left(\sum_{k=1}^{d_1}\sum_{l=1}^{d_2} c_{ik}^1 c_{jl}^2 h_{kl}\right) \left|\beta_i^{(1)}\right\rangle \otimes \left|\beta_j^{(2)}\right\rangle = 0$$

where $\left|\gamma_k^{(1)}\right\rangle = \sum_{i=1}^{d_1} c_{ik}^1 \left|\beta_i^{(1)}\right\rangle$, $\left|\gamma_l^{(2)}\right\rangle = \sum_{j=1}^{d_2} c_{jl}^2 \left|\beta_j^{(2)}\right\rangle$, and the symbols c_{ik}^1 and c_{jl}^2 denote the appropriate expansion coefficients. The linear independence of the set B implies

$$\sum_{k=1}^{d_1}\sum_{l=1}^{d_2} c_{ik}^1 c_{jl}^2 h_{kl} = 0$$

Now since B^1, C^1 and B^2, C^2 are valid bases in V^1 and V^2, the matrices c_{ik}^1 and c_{jl}^2 are necessarily invertible. Multiplying by $\left(c^1\right)_{k'i}^{-1} = b_{k'i}^1$ and $\left(c^2\right)_{l'j}^{-1} = b_{l'j}^2$, and summing over i and j, the left hand side of the above equation yields

$$\sum_{k=1}^{d_1}\sum_{l=1}^{d_2} \left(\sum_{i=1}^{d_1} c_{ik}^1 b_{k'i}^1\right) \left(\sum_{j=1}^{d_2} c_{jl}^2 b_{l'j}^2\right) h_{kl} = \sum_{k=1}^{d_1}\sum_{l=1}^{d_2} \delta_{kk'} \delta_{ll'} h_{kl}$$

$$= h_{k'l'}$$

Since k' and l' are arbitrary, we have $h_{kl} = 0$ for arbitrary k and l.

Theorem (representation of tensor product vectors): The representation of a tensor product vector $|\phi\rangle = \left|\mu^{(1)}\right\rangle \otimes \left|\nu^{(2)}\right\rangle$ in $V^1 \otimes V^2$, in the basis $B = B^1 \otimes B^2$ where $B^1 = \left\{ \left| \alpha_i^{(1)} \right\rangle ; i = 1, 2, \ldots, d_1 \right\}$, $B^2 = \left\{ \left| \beta_j^{(2)} \right\rangle ; j = 1, 2, \ldots, d_2 \right\}$ are bases in V^1 and V^2 respectively, is given by

$$ |\phi\rangle = \sum_{i=1}^{d_1} \sum_{j=1}^{d_2} m_i n_j \left| \alpha_i^{(1)} \right\rangle \otimes \left| \beta_j^{(2)} \right\rangle $$

with $m_i, n_j \in \mathbb{C}$, and where the vectors $\left| \mu^{(1)} \right\rangle$ and $\left| \nu^{(2)} \right\rangle$ are represented in the B^1 and B^2 bases respectively as

$$ \left| \mu^{(1)} \right\rangle = \sum_{i=1}^{d_1} m_i \left| \alpha_i^{(1)} \right\rangle $$

$$ \left| \nu^{(2)} \right\rangle = \sum_{j=1}^{d_2} n_j \left| \beta_j^{(2)} \right\rangle $$

It is clear that an arbitrary vector in V cannot be expressed as a tensor product. However, an arbitrary vector can always be expressed as a linear combination of tensor product vectors.

Inner product

Theorem: If B^1 and B^2 are orthonormal bases in the inner product spaces V^1 and V^2, then the basis $B = B^1 \otimes B^2$ is an orthonormal basis in $V^1 \otimes V^2$.

Tensor product operators

Theorem: The tensor product $\hat{A}^{(1)} \otimes \hat{B}^{(2)}$ of two operators $\hat{A}^{(1)}$ and $\hat{B}^{(2)}$ is equal to the ordinary operator product $\hat{A}\hat{B}$ of their extensions.

Theorem: The extensions of two linear operators acting on V^1 and V^2 to $V^1 \otimes V^2$ always commute.

Theorem (representation of tensor product operators): If a linear operator $\hat{A}^{(1)}$ acting on V^1 is represented by the matrix $A^{(1)}$ in a basis $B^1 = \left\{ \left| \alpha_i^{(1)} \right\rangle ; i = 1, 2, \ldots, d_1 \right\}$, and if a linear operator $\hat{B}^{(2)}$ acting on V^2 is represented by the matrix $B^{(2)}$ in a basis $B^2 = \left\{ \left| \beta_j^{(2)} \right\rangle ; j = 1, 2, \ldots, d_2 \right\}$, then the *representation of the tensor product operator* $\hat{C} = \hat{A}^{(1)} \otimes \hat{B}^{(2)}$ in the basis $B = B^1 \otimes B^2$ is given by the tensor product matrix[16] $C = A^{(1)} \otimes B^{(2)}$:

$$ C_{i,k;j,l} = A_{ij}^{(1)} B_{kl}^{(2)} $$

---

[16]The rows and columns of a tensor product matrix are designated by a pair of indices.

Eigenvalues and eigenvectors

Theorem: If the operator $\hat{A}$ acting on $V^1 \otimes V^2$ is the extension of a linear operator $\hat{A}^{(1)}$ acting on V^1, then the spectra of $\hat{A}$ and $\hat{A}^{(1)}$ are identical.

Theorem: If the operator $\hat{A}$ acting on $V^1 \otimes V^2$ is the extension of a linear operator $\hat{A}^{(1)}$ acting on V^1, and if $\left| \alpha^{(1)} \right\rangle \in V^1$ is an eigenvector of $\hat{A}^{(1)}$ with eigenvalue λ, then $\left| \alpha^{(1)} \right\rangle \otimes \left| \beta^{(2)} \right\rangle$ is an eigenvector of $\hat{A}$ with the same eigenvalue λ for all $\left| \beta^{(2)} \right\rangle \in V^2$.

If the degree of degeneracy of an eigenvalue λ of $\hat{A}^{(1)}$ is g, then the degree of degeneracy of the eigenvalue λ of the extension $\hat{A}$ of $\hat{A}^{(1)}$ to $V^1 \otimes V^2$ is gd_2, where d_2 is the dimension of V^2.

Theorem: If $\hat{A}^{(1)}$ and $\hat{B}^{(2)}$ are linear operators acting on the vector spaces V^1 and V^2 respectively, and if $\left| \alpha^{(1)} \right\rangle \in V^1$ and $\left| \beta^{(2)} \right\rangle \in V^2$ are the respective eigenvectors of $\hat{A}^{(1)}$ and $\hat{B}^{(2)}$ with eigenvalues a and b, then $\left| \alpha^{(1)} \right\rangle \otimes \left| \beta^{(2)} \right\rangle$ is an eigenvector of $\hat{A} + \hat{B}$ with eigenvalue $a + b$, where $\hat{A}$ and $\hat{B}$ are the extensions of $\hat{A}^{(1)}$ and $\hat{B}^{(2)}$, respectively, to $V^1 \otimes V^2$.

Complete set of commuting operators

Theorem: If the set of linear operators $\left\{ \hat{A}_1^{(1)}, \hat{A}_2^{(1)}, \ldots, \hat{A}_r^{(1)} \right\}$ forms a CSCO in V^1, and if the set of operators $\left\{ \hat{B}_1^{(2)}, \hat{B}_2^{(2)}, \ldots, \hat{B}_s^{(2)} \right\}$ forms a CSCO in V^2, then the set $\left\{ \hat{A}_1, \hat{A}_2, \ldots, \hat{A}_r, \hat{B}_1, \hat{B}_2, \ldots, \hat{B}_s \right\}$ will form a CSCO in $V^1 \otimes V^2$, where $\hat{A}_i$ and $\hat{B}_j$ are the extensions of $\hat{A}_i^{(1)}$ and $\hat{B}_j^{(2)}$, respectively, to $V^1 \otimes V^2$.

Bibliography

M.L. Boas. *Mathematical Methods in the Physical Sciences, 3rd Ed.* John Wiley & Sons Inc., 1980.

M. Chester. *Primer of Quantum Mechanics.* John Wiley & Sons Inc., 1987.

C. Cohen-Tannoudji, B. Diu, and F. Laloë. *Quantum Mechanics, Vols I & II.* Hermann and John Wiley & Sons Inc., 1977.

P.A.M. Dirac. *The Principles of Quantum Mechanics, 4th Ed.* Oxford University Press, 1958.

R.P. Feynman, R.B. Leighton, and M. Sands. *The Feynman Lectures on Physics, Vol-III.* Addison-Wesley Publishing Company Inc., 1963.

D.J. Griffiths. *Introduction to Quantum Mechanics, 2nd Ed.* Pearson Education Inc., 2005.

P.R. Halmos. *Finite-Dimensional Vector Spaces, 2nd Ed.* Springer-Verlag Inc., 1987.

H. Margenau and G.M. Murphy. *The Mathematics of Physics and Chemistry, 2nd Ed.* D. Van Nostrand Company Inc., 1956.

Index

Printed in the United States
by Baker & Taylor Publisher Services